The Loc

John,
best wishes
Jim Peek

The Lochsa Elk Herd:

History and Future

By

James M. Peek, Thomas A. Leege
and Michael W. Schlegel

**Cambridge
Scholars**
Publishing

The Lochsa Elk Herd: History and Future

By James M. Peek, Thomas A. Leege and Michael W. Schlegel

This book first published 2020. The present binding first published 2023.

Cambridge Scholars Publishing

Lady Stephenson Library, Newcastle upon Tyne, NE6 2PA, UK

British Library Cataloguing in Publication Data
A catalogue record for this book is available from the British Library

ISBN (10): 1-5275-9259-6
ISBN (13): 978-1-5275-9259-9

TABLE OF CONTENTS

LIST OF FIGURES

List of Tables

ACKNOWLEDGEMENTS

Clark Bainbridge, Neil Darby, Mary Messaros Vales, and David Woodall performed the arduous and tedious task of compiling records concerning the Lochsa drainage from the Lewiston Tribune, U.S. Forest Service, Idaho Department of Fish and Game, and Plum Creek Timber Company. U.S. Forest Service personnel, in particular, Dan Davis, Margaret Kirkeminde, Karen Ragotski, Dave Schoen, and Dennis Talbert, provided us with their records of timber harvest, road construction, wildfire, prescribed fire, and maps. Pete Zager assembled records and reviewed drafts of the manuscript. We thank Plum Creek Timber Company for allowing us to dig through their timber harvest and road construction records to provide a complete coverage for the Lochsa River drainage. Troy Tady, University of Idaho, developed the spectral habitat classification model necessary for analysis of plant community change over time. Kang-tsung "Karl" Chang, formerly of the University of Idaho Geology Department, was instrumental in developing the initial estimates of plant community composition in the Lochsa. Jocelyn Aycrigg, University of Idaho, updated the vegetation composition in the Lochsa drainage, giving a current picture of changes towards the coniferous forest of today. Jocelyn also produced the map of the Lochsa River drainage and reviewed a draft. Albert Stage (1929-2008), U.S. Forest Service, developed the forest succession sequence that illustrates plant community changes following fire in the absence of disturbance. Jeffrey Yeo provided important comments and suggestions on numerous drafts of the manuscript. Win Green, former supervisor, Clearwater National Forest, facilitated interviews, provided information concerning the Lochsa drainage and reviewed chapters. Eric Barker delved into the records of the Lewiston Tribune to locate the date the old Selway Game Preserve was abolished. John Beecham and Jim Hayden provided the Idaho Department of Fish & Game records on bear and cougar harvest plus wolf numbers and take. Dennis Baird of the University of Idaho Library sleuthed out the historical record and has been an invaluable source of much of our materials. His location of the aerial photos taken by the Washington Air National Guard in 1935 for the U.S. Forest Service following the Pete King fire provide invaluable insight into how much elk habitat was produced following those fires. Aaron Miles, Natural Resource Manager, Nez Perce Tribe, provided comments on the history and contemporary views of the Tribe. Charles

McKetta flew Peek into the Lochsa drainage to photograph current vegetation conditions in summer 2016. George Hatley, "Mr. Appaloosa" (1924-2011), Don McPherson, Conservation Officer (1926-2011), Idaho Fish and Game, Herman Kuykendall (1935-2020) and Jim Renshaw, former outfitters, were interviewed to learn more of the history of the Lochsa elk. Gary R. (Sam) McNeil (1938-2016), regional wildlife manager headquartered at Lewiston, provided highly valuable insights into the recent history of elk management in the Lochsa. This work would not have proceeded without the contributions of these people. This work was supported by Rocky Mountain Elk Foundation, Idaho Department of Fish & Game, U.S. Forest Service, University of Idaho College of Natural Resources and Idaho Cooperative Fish & Wildlife Research Unit.

CHAPTER ONE

INTRODUCTION

Central Idaho requires travel from the prairies down into the river canyons, contrasting with most of the intermountain west where people travel in the valleys and look up into the mountains. But central Idaho with its steep canyons that were made by ancient rivers and glaciers, has been travelled as much along the higher country as in the valleys. When Lewis and Clark first entered what is now Idaho, they looked down on the Lochsa River. If you live on the Camas Prairie in Grangevillle or Cottonwood or one of the smaller towns, you drive down to the Clearwater River or to the Salmon River. If you live on the Palouse, you drive down to the confluence of the Clearwater and the Snake River at Lewiston and Clarkston. Of course, if you live at Kamiah or Kooskia or Orofino along the Clearwater, you look up to see the original inhabitants' view from their habitations. If you drive north towards Coeur d'Alene from the Palouse, you eventually come down onto Coeur d'Alene Lake. The geology of central Idaho is primarily derived from the sculpting by the great ancient rivers in the region, resulting in the remnants of those same rivers being surrounded by steep slopes and stream sides that were not easily accessible to original peoples travelling through, and require we today to drive in "up and down country".

When Lewis and Clark came across the Bitterroots in 1805 and again in 1806, they travelled the trail on the high ground used by the native peoples, the Lolo Trail. And going across to the west the first time, they travelled in a huge snow fall in September. There is little doubt that the hunters they had with them were skilled, but did they really scour the Lochsa drainage and locate game in such conditions? And when they finally got down to the Clearwater and contacted the Nez Perce people, were they just given the dried salmon and dogs to eat and the tribesmen kept those tasty elk for themselves? We know now that the tribesmen hunted elk in the Lochsa. Lieberg (1899) met Indians that were burning in the Lochsa, probably to stimulate berry-producers but perhaps elk habitats as well. Aboriginals across the continent are known to use fire for purposes of acquiring game and creating habitats for various plants and animals. Perhaps those tribesmen in 1805 didn't have any elk to give Lewis and Clark. Perhaps they didn't have enough elk but had plenty of salmon to give. Perhaps they had plenty of elk and were not inclined to share with those emaciated white intruders. Perhaps the salmon were the prized protein and not elk. Perhaps they were more interested in creating habitat for berry-producing shrubs and palatable forbs, with elk just being an added benefit. Likely, tribesmen were purposefully burning for all of these reasons, but why Lewis and Clark were provided salmon remains a mystery.

The stimulus for this review came from a field trip initiated by the Rocky Mountain Elk Foundation into the Fish Creek drainage in July 1990. Discussions about future management of the population and the habitat inevitably led to discussions about the past and what had been learned. A large body of information on elk and their habitat has accumulated over the past century for the Lochsa River drainage, from the Idaho Department of Fish & Game (IDFG), the United States Forest Service (USFS), the University of Idaho (UI), and from many individuals. This information is virtually a history of elk and elk habitat management in northern Idaho. It was apparent to all that the record available in project reports and separate publications would, if synthesized, prove useful in providing direction for the future.

Reviews of conservation and management of the Scottish red deer (Darling 1937), the red deer on Scotland's Island of Rhum (Clutton-Brock et al. 1982), the Sun River elk population in Montana (Picton and Picton 1975), the Gallatin elk in Montana (Lovaas 1968), the northern Yellowstone National Park elk (Houston 1982), the Jackson Hole elk in Wyoming (Boyce 1989), and the White River elk in Colorado (Boyd 1970, Freddy 1985) provide historical perspective on those populations which serve as sources from which managers and others have benefited. Changes in management direction as conditions changed, or better understanding of the ecology and behavior as information accumulated, are evident for each of these populations. Management in the Lochsa and elsewhere in North America, initially intended to increase populations gave way to concern over damage to habitat through grazing, then to effects of land use practices such as logging on summer ranges, then to concerns over how to spread harvest equitably among increasing numbers of hunters, to concerns over depletion of bulls in populations, and now to the relationship of predators to this species.

Our purpose was to accumulate, analyze, and synthesize the available records relevant to the management of the Lochsa elk herd. We attempt to place the population dynamics of the elk herd and the ecology of its habitat in context of other reviews of elk biology.

Information Sources Used in This Report

Available records and documents concerning elk population management from IDFG include checking station data, hunter report cards, telephone survey results, mandatory reports, hunter densities, and elk census and population characteristics. Additional information on elk movement patterns and food habits was obtained from IDFG and USFS reports, University of Idaho publications, and graduate student theses.

Logging, roads, prescribed burning, and wild fire data for the Lochsa drainage came from the Clearwater National Forest Supervisor's Office, Powell Ranger District, and Lochsa Ranger District. Plum Creek Timber Company, Missoula, supplied access to their logging and road records for their lands in the upper Lochsa drainage. The early surveys of elk were provided by the Clearwater National Forest and the Nezperce National Forest.

We used a geographic information system (GIS) to store, analyze, and display spatial data on elk, logging, roads, prescribed burns, and wild fires. The software for the GIS is PC ARC/INFO from the Environmental Systems Research Institute. The project included four categories of map data. The first shows compartment (drainage basin) boundaries digitized from 1:24,000 base maps. Roads were the second category of the map data and also were digitized from 1:24,000 U.S. Geological Survey base maps. Spatial and temporal distribution of wildfire and prescribed fires were digitized from 1:100,000 maps, obtained from the Powell and Lochsa ranger districts.

Attribute data consist of elk population, elk harvest, logging, and roads. The first three contain historical records based on compartments, each of parts of the drainage. Elk and logging data often have more than one record for a compartment. The road data include road number and year of construction.

We assessed vegetation changes over the past 20 years by analyzing multi-spectral scanner (MSS) satellite imagery taken in 1972 (the first year such images were obtained) and 1991. Landsat 5 Multispectral Scanner imagery for summer 2012 (ESRI ArcGIS 10.4.1) was used to classify habitat within the Lochsa River drainage, which was defined by the Game Management Unit 12 boundary, into the five habitat types. Spectral classes were assigned to one of the five vegetation categories based on models developed from over 130 reference points of known vegetation.

CHAPTER TWO

LOCHSA RIVER DRAINAGE

The Lochsa River is designated Game Management Unit 12 by Idaho Department of Fish and Game (Figure 2-1). The U.S. Forest Service subdivides Unit 12 into 63 compartments for management purposes. Bisected from the northeast to the southwest by the Lochsa River and U.S. Highway 12, the 789,114 acre (319,343 ha) area extends to the Idaho/Montana border and Lolo Pass in the east and nearly to the small community of Lowell along the Clearwater River in the west.

The southern portion of Unit 12 includes the northernmost portion of the Selway-Bitterroot Wilderness. A detailed description of the Unit 12 boundary is provided in the annual big game season hunting regulations and maps by IDFG. U.S. Forest Service, Clearwater-Nez Perce National Forests, provide a detailed map of the National Forests that include the Lochsa.

The area is steep, mountainous terrain, dissected by innumerable streams and draws (Figure 2-2). Elevations range from about 1312 feet (400 m) near Lowell to over 8530 feet (2,600 m) on peaks dotting the game unit's perimeter. Geology is complex, but the basic formation is the Idaho Batholith which is primarily composed of coarse-grained quartz (Greenwood and Morrison 1967). Western redcedar, grand fir, subalpine fir, Douglas fir, mountain hemlock, and western hemlock associations define the climax communities of much of the area but elevation, aspect, and successional stage define current vegetation. Ponderosa pine and Douglas fir associations prevail at lower elevations and on drier sites. Subalpine fir, lodgepole pine, and Engelmann spruce associations dominate higher elevations. Seral shrubfields, Ponderosa pine, Douglas fir, western white pine, and lodgepole pine occur following disturbance of these forests, originally likely by various pathogens such as mountain pine beetles or spruce budworm that weakened and killed trees, which subsequently became victims of windfall or blowdown, and finally by wildfire. In recent times logging is also a disturbance, most recently in the eastern portions of the drainage on the Powell Ranger District and the private "checkerboard" sections that were originally given to the Northern Pacific Railroad and subsequently to its

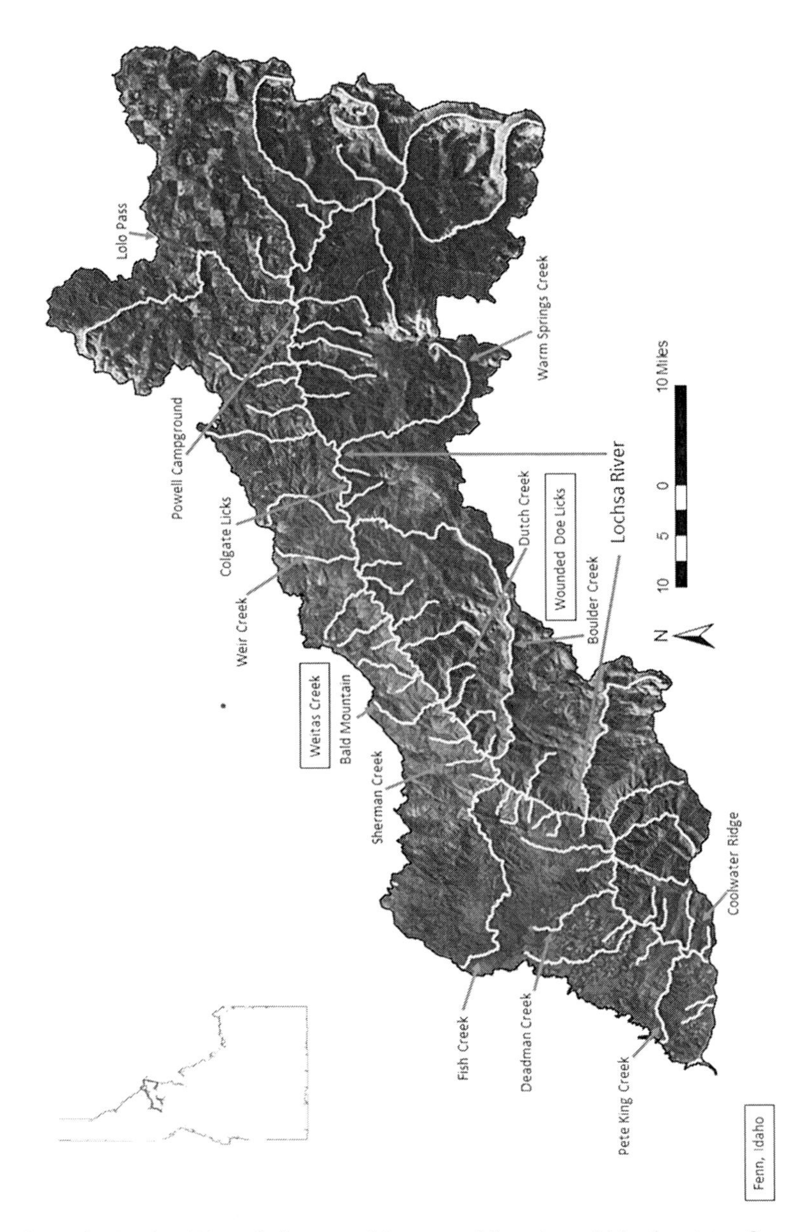

Figure 2-1. The Lochsa River drainage and its general location within the state of Idaho.

Figure 2-2. Photo of Lochsa River facing north, taken in 1980 by U.S. Forest Service. U.S. Highway 12 is visible along the River. Drainage to left (west) is Holly Creek, Skookum Creek to right (east) in the high foreground. This is approximately 80 km east of the confluence of the Lochsa River and the Selway River. Darkest color is mature timber, lighter is regenerating timber, and lightest is nonforested area. Courtesy of University of Idaho Library.

land management spinoff, Plum Creek Timber Company, and most recently Western Pacific Timber. Current status of the private timberlands is in flux as Western Pacific Timber seeks to divest itself of those holdings.

The Lochsa drainage experiences a climate that is "transitional between a north-Pacific coastal type and a continental type" (Finklin 1983:11). Temperatures are moderate compared to similar latitudes further east. Average daily temperatures may range from 28^0 F (-2 C) in January to 89^0F (21 C) in July. Peaks of precipitation occur during fall and winter with a secondary peak in late spring. Annual total precipitation may range from 40-70 inches (1.02-1.78 meters) depending on elevation. During winter deep snows often occur although snow depths are mediated by elevation and aspect. Daily weather records have been recorded nearby at the Fenn Ranger Station on the lower Selway River since 1939 (Table 2-1). This station is located about five miles up the Selway River from its confluence with the

Table 2-1. Climate summary for Fenn Ranger Station, 1939-2016. Location is about 5 miles (8 km) from the confluence of the Selway and Lochsa Rivers on the Selway [1].

	Jan	Feb	Mar	Apr	May	Jun	Jul	Aug	Sep	Oct	Nov	Dec	Total
Average Maximum Temperature (C)	1.8	6.1	10.8	16.3	21.4	25.4	31.7	31.6	24.7	15.9	7.1	2.7	16.3
Average Minimum Temperature (C)	-4.6	-2.7	-0.8	1.9	5.4	8.8	10.9	10.2	6.9	2.9	-0.7	-3.4	2.9
Average Total Precipitation (cm.)	8.8	8.6	9.4	9.1	8.9	8.4	2.5	3.0	5.3	7.4	11.4	10.9	96.8
Average Total SnowFall (cm.)	43.2	20.6	8.4	0.5	0.0	0.0	0.0	0.0	0.0	0.3	11.4	26.6	120.9
Average Snow Depth (cm.)	17.8	12.7	2.5	0.0	0.0	0.0	0.0	0.0	0.0	0.0	0.0	7.6	45.6

[1]Data from Western Regional Climate Center, Reno, Nevada.

Lochsa at 1560 feet (475.5 meters) elevation. There are disjunct species found on the west coast such as Pacific dogwood and scotch broom, reflecting the high moisture regime and the moderate temperatures in the lower area. Annual snowfall average is 47.6 inches (1.21 meters) with most of that occurring from December through February, although snows are also recorded in October, November and March. Average snow depth is 7 inches in January, 5 inches in February and 1-3 inches (0.03-0.09 m) in April and December, respectively.

The Powell Ranger Station records illustrate the climate at the upper elevations at 5873 feet (1790 meters) near Lolo Pass (Table 2-2). Annual temperatures range between 55.9 F (13.3 C) average maximum for all months to 29.5 F(-1.39 C) average minimum for all months. Temperatures for January and December average at freezing for the highs and in the low 20's and high teens for the lows. Total snowfall averages 166.5 inches (4.23 meters), with depths averaging between 2 and 2.7 feet (0.6-0.8 meters) in January through March. The Lochsa drainage climate can be characterized as wet and mild in comparison to most of the northern Rocky Mountain forests.

Access and Timber Harvest

Road construction started in 1926 in the Lochsa by the USFS. The highway up the Lochsa was completed as a gravel road in 1960. Two years later, 1962, the road was paved. Two peak years of road building occurred: 1935 when 145 miles (234 km) of road were built which followed the extensive fires of 1934 and coincided with the presence of the Civilian Conservation Corp, and 1969 when 178 miles (286 km) of road were built which coincided with sharply increased timber harvest (Figure 2-3). Roads are concentrated in the east and west ends of the drainage with few roads south of the Lochsa River. Road building averaged 19±-30 miles (30 ± 49 km) built annually between 1926 and 1990. Total miles of road built during the 65-year period equaled 1194 miles (1,922 km).

Our records for timber harvest begin in 1950. The area of timber harvested annually increased in general through 1990 with precipitous decline after that (Figure 2-3). Similar to trends throughout the Northwest, peak logging occurred in the 1980s. Between the 1940s and 1992, 57,351 acres (23,212 ha) of the drainage have been logged. As would be expected, the distribution of logging coincides with the distribution of roads that are concentrated at the east and west ends of the Lochsa drainage with little occurring on the south side of the Lochsa River.

Table 2-2. Climate summary for Powell Ranger Station near head of Lochsa River and Lolo Pass[1]. Data is for the period August 1962- July 2013.

	Jan	Feb	Mar	Apr	May	Jun	Jul	Aug	Sep	Oct	Nov	Dec	Total
Average Max. Temperature (C)	0.1	3.6	7.6	12.1	17.8	22.4	28.2	27.9	22.1	13.9	4.3	-0.5	13.3
Average Min. Temperature (C)	-8.9	-7.9	-5.1	-2.1	1.2	5.2	7.2	6.2	2.2	-1.5	-4.6	-8.7	-1.4
Average Total Precipitation (cm.)	14.5	9.4	8.4	6.9	7.6	7.6	3.3	3.8	5.6	7.4	11.7	12.7	98.3
Average Total SnowFall (cm.)	124.	70.4	51.3	17.3	2.8	0.3	0.0	0.0	0.0	4.1	45.7	107.	424.
Average Snow Depth (cm.)	71.1	81.3	60.9	15.2	0.0	0.0	0.0	0.0	0.0	0.0	7.6	35.6	22.9

Roads and Timber

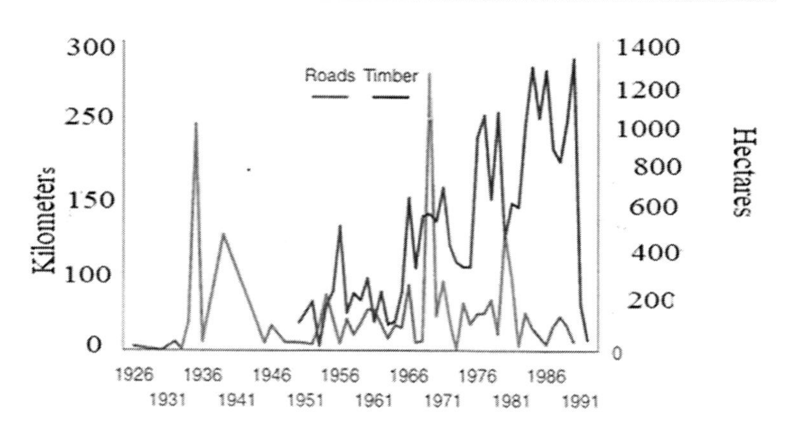

Roads and Timber

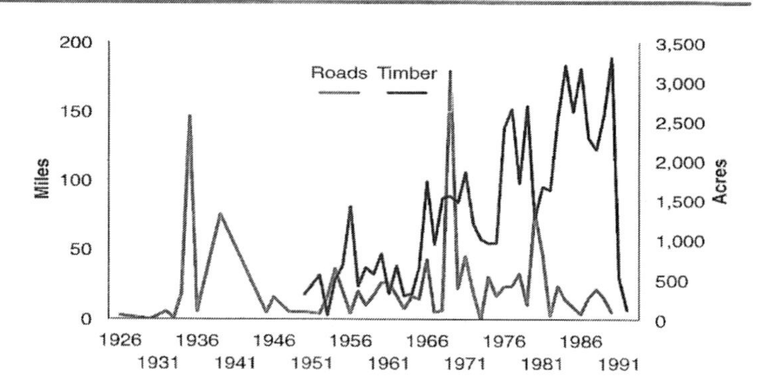

Figure 2-3. Length of roads built, and area of timber cut in the Lochsa from 1926-1991. This includes Clearwater National Forest and Plum Creek Timber harvests.

Some logging and associated road construction has occurred since1991, but this shows the trends. More recent activity is at higher elevations on the eastern portions of the drainage. These cuts are not much used by elk because of deep snows and the presence of wolves. These activities illustrate efforts by U.S. Forest Service to proceed with multiple use management as much as possible. Campgrounds along the Lochsa River are also available, with the DeVoto Memorial Grove with its huge western red cedars being one of the more attractive sites. The River attracts floaters, kayakers, and fishermen from late spring through fall. A number of suspension bridges provide access for hikers and stock to cross the river to the south to the Selway-Bitterroot Wilderness. Roads and trails provide access to the north side of the river, but some roads are managed to reduce vehicle access that affects elk distributions.

CHAPTER THREE

ORIGINS OF ELK IN THE LOCHSA

Elk are members of a large group of antlered species in the genus *Cervus* that occur across the northern hemisphere. Wherever their relatives and they themselves occur, they are and have always been a prize for their flesh, hides, antlers, teeth, and bones. They represent a highly adaptable complex of species and even today the taxonomy of the complex is not completely agreed upon. As one would imagine, taxonomic status of North American elk, more properly designated wapiti to distinguish it from the European elk (North American moose), has been examined and re-examined numerous times, as indicated by changes in its scientific name. We use elk and wapiti interchangeably throughout this report. It was originally designated by Carl von Linnaeus as *Cervus elaphus* in *1758*, part of the red deer complex that occurs across the northern hemisphere. In 1780, attention to what the wapiti should be designated finally resulted in classifying it as *C. canadensis*. However, after extensive studies of behavior, distribution, and morphology, the scientific name was changed back to *C. elaphus* in 1973 (Jones 1983). Recent analyses suggest that the European and Asian red deer complex, plus the North American wapiti, may well be three different species (Geist 1998, O'Gara 2002, Polziehn and Strobeck 1998). The international journals are now using *C.canadensis so* the issue is likely settled for now and our wapiti is classified as it was prior to 1973.

From the standpoint of management and conservation of wapiti in the western part of North America, the scientific designation at the species level doesn't matter as much as the designation of subspecies. The Tule elk, Roosevelt elk, Manitoba elk, and Rocky Mountain elk are considered different subspecies, with management significance. Conservation of Tule elk currently includes introductions to formerly occupied range and otherwise suitable habitat in California. Roosevelt elk occur west of the Cascade Range in Canada and the US. Rocky Mountain elk are thought to have been re-established to formerly occupied range primarily from introductions from Yellowstone National Park in the U.S. and Banff National Park in Canada (Polziehn et al. 1998). They now range east of the

Cascades clear to suitable habitat in states and provinces that border the Atlantic Ocean.

Introductions of Yellowstone elk into Idaho, including the Lochsa, occurred from 1915 through 1946 (O'Gara and Dundas 2002), with an additional 68 from eastern Oregon put into the Pete King drainage of the lower Lochsa in 1976 (Leege 1979). Recent analyses suggest that Lochsa elk are part of the elk population that extends across Idaho and adjacent states and provinces, owing to interchange between groups across this broad region. However, the Lochsa population appears to be moderately distinguishable within this huge complex (Aycrigg and Garton 2014), perhaps because of being less likely to move or intermingle with populations in other areas, or perhaps having less gene flow with the Yellowstone introductions than other populations in Idaho had.

Elk probably evolved in open country where grasses were important in the diet (Guthrie 1966, 1968). The grasses occurred in the drier forests as well as shrub/steppe and prairies. Their premolars and molars show low wide crowns (selenodont), adapted to grinding the silicaceous opaline phytoliths in grasses of the more open regions further east and south. The adaptation includes delayed attainment of complete adult dentition to about three years. Adult tooth wear is relatively fast when compared to the browse-feeding moose. A moose, which replaces all of its milk teeth at 21 months, has high crowned (hypsodont) premolars and molars adapted for cutting and crushing the lignaceous woody forage that is actually less wearing on dentition than grass. Both elk and moose show equivalent tooth wear patterns at the 8 to 10 year age range, but the moose has used its adult teeth approximately a year and a half longer than the elk (Peterson 1955, Quimby and Gaab 1957).

Elk are highly gregarious, a further adaptation to living in open habitat (Geist 1971). Herds of over 200 in summer and over 500 in winter may occur in more open country, although evidence from the northern Yellowstone suggests that those large herds that occupied the open slopes are readily detected by wolves and have largely disappeared with the advent of the gray wolf (Mao et al. 2005). The gregarious-ness facilitates early detection of predators, but wolves also tend to prey on elk in larger groups (Hebblewhite and Pletscher 2002).

Additionally, suitable forage in montane and prairie rangeland is usually more evenly distributed than in forests, where forage may be more dispersed and patchy. Group sizes in forested habitats, including harem size, are smaller. In forests, depending upon size and tree densities, the probability of seeing predators at a distance is lower as is the probability of predators seeing elk. Scent and sound also weigh in on these probabilities for both predator and prey and how this differs between forests and open country would be difficult to generalize. Gower et al. (2009) concluded that group sizes for any given elk population represent an interaction between forage availability, competition between individuals for food, and predation risk. While larger group sizes should occur where predation risk is high, Hebblewhite and Pletscher (2002) found elk living in small groups may rarely encounter wolves while predation increased as elk density increased. Variation in different areas suggest that generalizations were difficult.

There is evidence of the genus Cervidae, from which wapiti descended, in the early North American fossil record (Kurten and Anderson 1980, O'Gara and Dundas 2002). The record is unevenly distributed across the northwestern states and adjacent Canada but it does give an indication of early presence. The early relatives of wapiti first appeared in southwest Idaho during the late Pliocene epoch, around 2.4 million years ago (Kurten and Anderson 1980). The current-day wapiti appeared to be well distributed across the northwestern states and adjacent Canada as early as 8,000-10,000 years ago. They persisted in North America in glacial refugia, colonizing suitable habitat that became available as the continental ice sheet retreated northward (Geist 1971). Wapiti were present in the Yellowstone at least 2000 years ago, based on evidence from the Lamar Cave (Hadly 1990). The mountain-valley complex of northwestern Montana contained wapiti as far back as 4000-8000 years before present.

The excavations at Medicine Hat, Alberta, revealed the existence of wapiti along with dire wolves, gray wolves, a panther, mammoths, horses, moose, antelope, bighorn sheep, caribou and bison (Woodburne 2004). We live with many fewer large mammals than existed in Pleistocene times over 10,000 years ago.

Impressions of the Lochsa elk are bound up with the Lewis and Clark expedition and major catastrophic wildfire events in the first third of the twentieth century. Visions of large bands of elk trailing through shrub fields in the deep, heavy snows (Figure 3-1) are mixed in with the knowledge that the Lewis and Clark expedition found little sign of these animals when they passed through the region a century and a half before. The old perception that elk were primarily prairie animals and late arrivals to the mountains following the wildfires was undoubtedly spurred by these early events in northern Idaho.

However, Space (1964) reported that the Lewis and Clark journals did present evidence of elk in the Lochsa in 1805. Nez Perce guides reported elk plentiful in the Papoose Creek (renamed to Imnamatnoon Creek) area and around the Jerry Johnson and Colgate Hot Springs. Indians were said to hunt elk around those hot springs.

Robbyn Johnson, an archaeologist for the Clearwater National Forest at Kamiah, reported elk remains at the mouth of Boulder Creek and at Powell that were at least 3000 years old (personal communication to James Peek, 12 October 2000). These hunters were using atlatls, an open country weapon that would be suitable for use further west along the breaklands of the Clearwater. They sought to convert the forests to more open lands with fire, which grew edible plants for humans and wildlife alike. Johnson thought that the meadows and other open sites were probably continuously occupied for 5-6000 years, which suggests that harvests were not high enough at any time to indicate excessive harvest of the wildlife resource or the edible plants.

Additional evidence of elk being present in the Lochsa prior to 1800 was provided by Sappington and Carley (1989). Investigations of sites along the highway proposed for improvement required examination for artifacts that indicated the presence of aboriginals. Two sites were carefully examined by these archeologists, the Beaver Flat site and the Pete King Creek site. Beaver Flat, 4.5 miles west of the Boulder Creek campground, was initially occupied approximately 8,000 to 10,000 years ago. Bone fragments were present but were not identifiable. The Pete King site, near the confluence of the Selway and Lochsa Rivers, yielded numerous prehistoric artifacts. It was judged on the basis of the presence of hunting toolkits, and evidence of processing of elk, deer, and bear, that the site was used in fall and spring for hunting. Radio-carbon dating suggested occupation spanned from 2800-3100 years ago. Historical artifacts located in upper strata of this site included spent cartridge cases. Examination of sites distributed along the Clearwater River from the confluence of the Selway and Lochsa Rivers to its confluence with the Snake River show elk, deer, and bighorn sheep being present (Sappington 1994). Bison remains probably came from trips into Montana for taking that species. The archeological records and early accounts suggest that elk were present long before Europeans entered the region.

John Mullan's trip to survey the Lolo Trail in 1854 found elk signs east of Lolo Pass on Lolo Creek (Baird and Baird 2003). Their description of the Lolo Trail at Bald Mountain indicates extensive communities dominated by bear grass and mountain ash, similar to what Lewis and Clark recorded half a century earlier.

H.M. McCartney reported his trip of 1879 to locate a railroad route through the Bitterroots, going up the Lochsa drainage (Baird and Baird 2003). He recorded seeing over 200 elk at Elk Prairie, 150 miles east of Lewiston. McCartney wrote that this was the finest game country he had ever travelled in. The country was considered too rough for construction of a railroad.

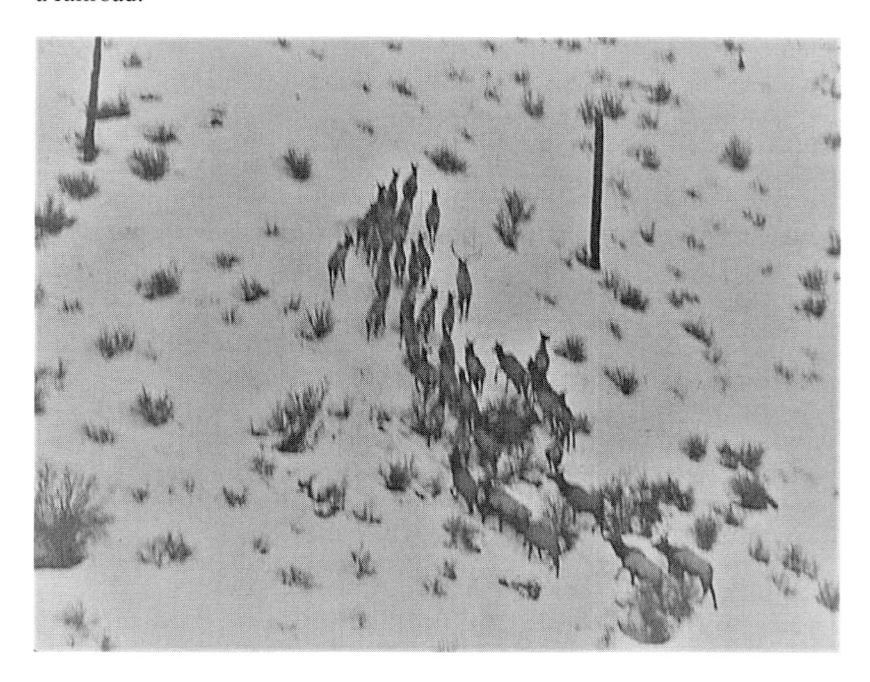

Figure 3-1. A band of elk on a shrub field in the Lochsa. Photograph by Tom Leege.

The surveys conducted in 1897 by Leiberg (1899, Figure 3-2,) reported that Indians burned approximately 1,110,000 acres over a 200 year period. One purpose of their trip was to hunt in the Lochsa and elk were definitely a part of the quarry.

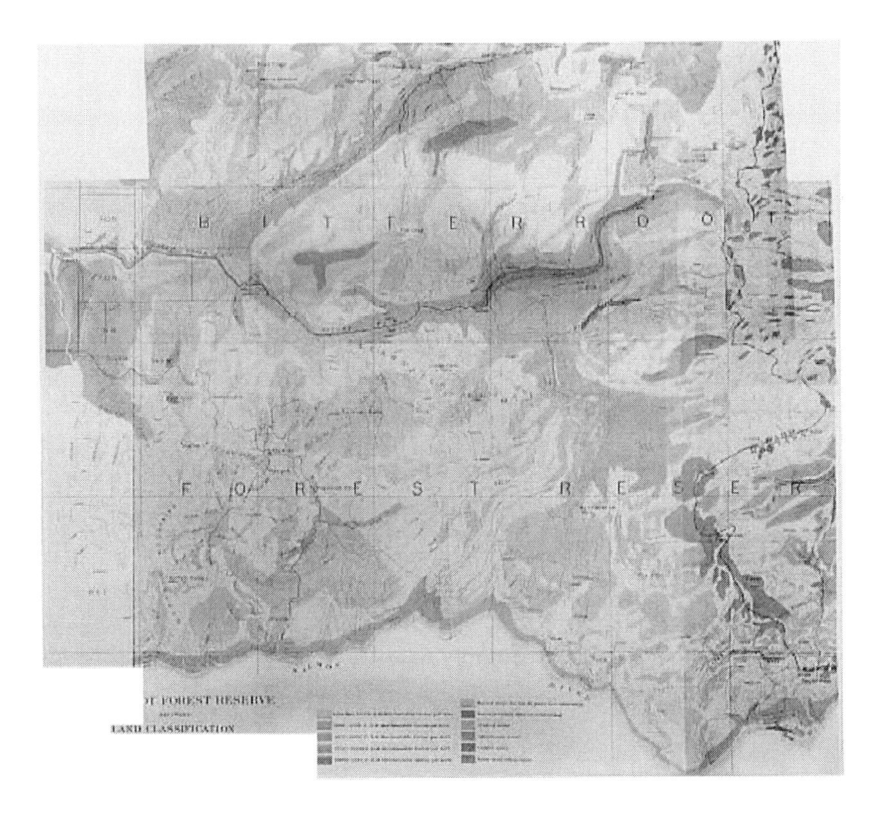

Figure 3-2. Lieberg surveys of 1899. Green areas represent forest, yellow represents regenerating forest, and red represents shrublands and other areas that are nonforested. The uppermost green is the cedar stands along the Lochsa River. The middle green is the Selway River with the darkest green being Moose Creek. The thin lowest green is the Salmon River. *See centrefold for this image in colour.*

***

Northern Idaho forests evolved in the presence of catastrophic wildfire, and most stands present today that have not been logged developed following a burn (Wellner 1970). In a region where deep snows preclude access to ground-level forage, the fields of tall shrubs, will develop within a few years after fires replace the forest overstory. Many of these shrubs provide an abundant and palatable source of forage for animals that can negotiate the deep snows found in the shrub fields. While further north in British Columbia, moose were able to proliferate on newly created browse ranges following fires in this century (Hatter 1950), elk were the primary beneficiaries of this newly created habitat in northern Idaho and adjacent states. Quite likely, over the time since these coniferous forests were established following the recession of the glaciers, and establishment of the weather pattern of hot and dry summers with thunderstorms that caused fire, elk populations ebbed and flowed depending upon condition and amount of suitable habitat .

Early presence of elk in the Lochsa may be attributable to a number of factors. They may have occurred early on in isolated patches of suitable habitat following the glacial retreat. Vegetation immediately following the glaciers would have been primarily herbaceous and/or low-growing shrubs rather than conifer forest. Elk may have been able to occupy these early post-glacial areas and subsequently may have been more or less isolated in suitable areas as the post-glacial coniferous forest developed. Subsequently, populations may have been expected to proliferate as the climate that allowed wildfire to occur developed. Under these conditions, predation may have been at levels that allowed elk to perpetuate themselves.

The higher elevations of the Lochsa drainage, above 5500 feet (1,675 m), consist of subalpine fir/Engelmann spruce habitat types interspersed frequently with mountain meadows and dry parks, constituting relatively high-quality summer-early fall range (Hash 1973). Similar vegetation was also present in early post-glacial times (Brunelle et al 2005). These areas may have been occupied by elk which migrated to winter range east of the Bitterroot divide or south to drier sites in the adjacent Selway River drainage, or perhaps west to the Middle Fork of the Clearwater. This seems consistent with evidence that elk occurred in low density further west in the lower Snake River, as indicated by presence of their remains in middens, further suggesting their presence throughout the inland Northwest at varying distributions and densities.

While major fires apparently occurred in the 1700's in the Lochsa (Lieberg 1899), and again in 1830 (Space 1964), the scenario suggests that no major elk population increases occurred after those early fires. However, the large increases in elk did follow the extensive fires that occurred in the mid-1900's. Human exploitation coupled with predation by wolves, cougars, and bears, may have been sufficient to keep elk populations at low levels regardless of habitat conditions prior to the 1900's. That the increase in elk in the mid-1900's following the big fires may be attributable as much to the release from extensive predation and human exploitation as to habitat conditions cannot be discounted. If smaller populations of elk using isolated and patchy habitat in winter were the prevailing scenario prior to 1910, then predation could conceivably have limited population size. This possibility is given more credence by the increase of elk following reductions in predators in the early decades of the 20th century, plus the increase in numbers following elimination of antlerless hunting in 1975 even as deterioration of the winter shrub-field forage supplies continued. Limitation of ungulate prey by intensive predation has been extensively documented for caribou and moose in Canada and Alaska (Bergerud and Elliott 1986, Gasaway et al. (1983,1992) and elk in the Yellowstone (MacNulty et al. 2016, Garrott et al. 2009). Urness (1989) suggested that bison may have been limited west of the Rocky Mountains by these factors, with entirely suitable and extensive habitat being available.

These possibilities concerning presence and sizes of prehistoric elk populations in the Lochsa are tentative, since they are rooted in spotty information and an obviously imperfect understanding of the roles of mortality causes and habitat relationships of these elk back then. However, they are important hypotheses because of their implications for sustained management of the Lochsa elk and other populations occupying similar habitats. Today, with emphasis on restoring anadromous fish, reduced fires, presence of all predators except the grizzly, and concerns for other species, plus the continually declining forage supplies, elk will be present but not at the levels of the mid-1900's. Still, it is important to recognize that the forests of this country will also change just as has been the case in the past.

CHAPTER FOUR

ELK POPULATION BIOLOGY

The purpose of this chapter is to report some of the population biology that has relevance to the work done on the Lochsa elk. Factors that influence survival and mortality of the various age and sex classes are included. A fundamental principle of population biology is that groups of living organisms do not continue to grow forever, but do reach a level where growth ceases. Such is the case with elk, but reasons for cessation of population growth vary in time and space. While adult survival and age distribution will greatly influence population growth, calf survival typically fluctuates more between years and will generally have more influence on population size (Nelson and Peek 1982, Gaillard et al. 1998, Raithel et al. 2007, Lukacs et al. 2018). Circumstances where adult cow survival will come into play includes establishment of hunting seasons for either sex or cows to manage population levels. Also, extended severe winters can increase adult mortality when populations are at high levels. Populations with older age structures will be more vulnerable to abrupt declines when severe winters occur. Bulls are more susceptible to dying during severe winters, particularly when the winter starts early. Bulls will come out of the rut that occurs in late September into early October, in poor condition associated with rutting activity. Prolonged summer drought that reduces forage quality and prolonged winter snow depths will reduce calf survival. Populations at high densities relative to available forage are also subject to more mortality than populations that are at lower levels where the forage base is more plentiful for each individual. Predation is implicated in survival of all age and sex classes.

Typicaly in the larger herbivorous species, production and survival of calves is influenced by population density relative to available forage, age of females, whether females had a calf the previous year, and dominance rank of individual females (Clutton-Brock et al. 1987, Gaillard et al. 2000). Generally, middle-aged females of high dominance rank that have not produced the previous year are the most productive and survive severe winters and late summer drought better than other sexes and age classes. First year breeders, many of which will be two or three-year-olds, will be less productive than individuals that have produced young the previous year. Older individuals may well produce fewer calves, be more vulnerable to predators than younger adults, and have lower survival rates than younger individuals.

Two long-term investigations of *Cervus elaphus/canadensis* provide insight into reproductive performance of elk populations considered to be resource limited, that is, at or near a forage-based carrying capacity. The high reproductive rates of palpated elk in the Lochsa were comparable to those found in the northern Yellowstone by Greer (1966) and Houston (1982). Both populations have experienced severe winter die-offs. These elk populations, even when at levels that are forage-limited, show adequate levels of nutrition and are in good condition during most years. However, the Yellowstone elk respond to different levels and seasons of precipitation and exhibit lower levels of population growth when population levels are high (Houston 1982, Coughenour and Singer 1996). Thus the interaction between population density and forage availability will affect elk condition, and consequently interpretation of studies on condition. This concept of a compensatory response in calf production/survival when a population is at lower density is typical for most species, although species that reproduce at longer intervals such as grizzly bears may not show much of a response when compared to species that reproduce more frequently. The Northern Yellowstone herd was predicted to fluctuate about a mean of 14,000 (range 10,000- 18000) elk in the absence of significant hunter harvest and significant predation (Houston 1982). While severe winters and following springs are the times when elk mortality is highest, precipitation during spring which includes rain-on-snow events that preclude access to forage in the Yellowstone is related to survival. In the Lochsa, where forage is primarily browse that extends above the snow level, depletion of reserves related to heavy browsing pressure on palatable shrubs plus deep snows that cause increased energy expenditure are responsible.

The long-term records for the Northern Yellowstone elk population provide evidence that deep winter snows and high spring precipitation can reduce adult survival, and calf and bull mortality is related to population density (Coughenour and Singer 1996). These climatic conditions affect Lochsa elk as well. However, these studies were accomplished in the absence of the wolf. Predators of wapiti include black and grizzly bears, cougars, wolves, and coyotes, and occasionally golden eagles (Griffin et al. 2011, Schlegel 1986).

The work on Scotland's red deer is long-term and comprehensive, and likely applies to their North American cousins to a large degree. The red deer under study were isolated on the Island of Rhum off the Scottish mainland, and existed without predators. They are also smaller than our wapiti. This isolated population ranged between 57 and 166 individuals, was not hunted and was considered resource-limited at the higher densities.

Age, reproductive status, group size, dominance rank, and population size were investigated (Clutton Brock 1984, Clutton-Brock et al 1987). Higher proportions of middle aged hinds (7-13 years) produced calves with higher probability of survival that did younger and older hinds. Calves born with weights over 14 lbs were more likely to survive the summer than lighter-weight calves. More hinds that did not rear a calf the previous year had calves than those that had a calf. However, those hinds that did have a calf the previous year were more successful in bringing off a calf through the summer of the next year, likely because whatever was causing calf mortality operated in successive years. Also, hinds belonging to smaller groups were more successful in bringing off calves than those in larger groups. Clutton-Brock et al. (1987) reported that red deer fecundity at rising density was affected more in three-year-old females that were first breeders than in older females that were not lactating. Females that exhibited high dominance were also more fecund. As with the Yellowstone elk prior to the advent of the wolf, summer calf survival did not vary consistently with population density. The final conclusions in the Rhum investigations were that competition between individual groups at low densities may be in the form of scramble competition between groups where smaller groups may have access to more forage per individual than larger groups, while contest competition for forage between individuals in populations at high density is more prevalent.

Cook et al. (1996, 2001a, 2001b, 2004) provided evidence from observations of conditioned elk that late summer forage had a significant effect on condition of cows going into the rut and winter. These studies involved elk that could be followed closely and were otherwise tolerant of humans. Prolonged summer drought can reduce nutrient levels in forage plants enough to adversely affect the condition of lactating cows. This finding was a major change from traditional views that winter/spring forage availability was the major influence on production and survival. The work was preceded by investigations of calves born to captive cows by Thorne et al. (1976). Calves weighing 35 pounds or more at birth had a 90% chance of surviving to 4 weeks while calves weighing 25 pounds or less at birth had less than a 50% chance of survival. Typical birth weights range from 30 to 40 pounds (Johnson 1951, Schlegel 1976, Thorne et al. 1976) with birth weights below 30 pounds making calves more likely to die. In the wild, elk that are in poorer condition or younger individuals that are least able to escape the predator will be more vulnerable to predation. The predator-prey relationship is complex because many other factors are involved, including habitat relationships, population density in relation to the forage base, late summer forage conditions, and winter severity.

The Lochsa elk population was primarily monitored in winter to assess population size, sex, and age, plus harvest levels. The attributes discussed would all weigh in on these observations even if they were not identified during the more general surveys. Several investigations reported later did identify some of the attributes, but the routine monitoring for management purposes didn't have the details because they take more time and resources to identify.

CHAPTER FIVE

EARLY POPULATION INVESTIGATIONS

Investigations of the Lochsa elk population can be separated into the 1930-1958 period, and the decades from 1960 to the present. Earlier investigations were done from the ground in winter, with only segments of the Lochsa River system involved. The first surveys using aircraft were initiated during this earlier period with low reliability, but they did provide experience as to what was needed to do the initial helicopter surveys of the late 1950s. Fixed-wing aircraft were not especially useful given the methods used and the steep terrain that characterizes the drainage. The helicopter became the most successful type of aircraft for census and classification purposes, and has been used in most such work in Idaho since.

The U.S. Forest Service initiated winter game studies in the Selway and Lochsa drainages in 1934-35, with additional surveys conducted the next two winters as well. The purpose of these studies was to determine the number of deer and elk wintering in the Selway and Lochsa drainages, to determine the drift or migrations of game into the area from outside, to determine the amount of forage used and the influence of snow depth upon this use, to determine the carrying capacity of the winter range, and finally to determine the game losses and causes of these losses on this winter range (Landall and Fager 1935).

These surveys were done from the ground, and the reports detailed their conduct. Patrols were initiated in December 1934 and ended in February 1935, and again from October 1935-April 1936. As an example of the organization of the patrols, the 1935-36 patrol on the Clearwater National Forest section of the river was carried out by "two E.R.A. relief men and two E. R. A. ten-percenters, with three ten-percenters on the Lochsa District. Due to a shortage of funds the ten-percenters were released from December 23, 1935, to January 23, 1936" (Eddy and Rauma 1936). (E. R. A. was the Emergency Relief Administration). E.R.A. people were paid $44.00 per month with $0.42 per diem per day. The winter patrols were made on snowshoes by two, two-man crews stationed at the Lochsa Ranger Station at the mouth of Fish Creek and by 5 E.R.A. men plus two permanent employees stationed at Pete King Ranger Station. Note that the Lochsa

River at that time was administered by Nez Perce National Forest from the mouth to Snowshoe Creek, by the Clearwater National Forest from Snowshoe Creek to above Lake Creek, and by the Bitterroot National Forest above Lake Creek.

Snow conditions were documented by the patrols. The 1934-35 winter was considered mild. However, February 1936 snow depths at Lochsa Ranger Station were 35 inches, and the winter was considered severe from late January through mid-March. A shortage of food was noted and by mid-February elk had started to move down river, with an estimated 300 head moving onto the Nez Perce National Forest in the Bimerick and Deadman Creek drainages. Malnutrition loss was observed along with heavy tick infestations and parasites. The 1934 Pete King fire had burned over much of the lower Lochsa and the resulting temporary food shortage, made more serious by the deep snows, was considered the reason for the poor condition and malnutrition of the elk. The 1936-37 winter was again mild, with lower snow depths and not much cold weather. Totals of 1,064, 1,627, and 3,158 elk were counted on the Lochsa in 1934-35, 35-36, and 36-37 by these surveys (Table 5-1).

The classifications reported during these surveys do not coincide with the total estimated numbers (Tables 5-1 and 5-2). The total observed number in 1934-35 of 1064 was 423 elk higher than were reported as classified. The 1935-36 severe winter counts were equal to the number classified, which suggests that substantial numbers of elk were forced down to where they could be counted and counts in milder winters did not observe as high a proportion of the population and classifications were more difficult to make. Similarly, the estimates in Table 5-2 do not match the distributions reported in Table 5-3 because of different portions of the drainage that are included. However, the cow:calf ratios derived from the ground counts did not reach higher than 100:45 except for the small sample observed in 1936-37. If the cow:calf ratios do represent an approximation of herd composition, then they weren't high even when predators had been reduced. These surveys were discontinued during the war years and the next winter survey was made from January 19-31, 1948, by USFS and IDFG personnel. An additional check was made with a Stinson Voyageur plane on February 6th, which was reported thus " (Clearwater National Forest 1948):

> "A word about the adaptability of planes to census work. As far as actual counting is concerned, the present plane offers very little promise. You just cannot obtain a reliable count from the air."

Table 5-1. Summary of elk surveys, 1934-1950, in Lochsa River, not including the Powell District. Data taken from files of the Clearwater National Forest and Nez Perce National Forest.

Year	Ground[1] Seen	Air	Total	Est. Additional	Est. Total	Jan.-Feb. Snow Depth(cm)
1934-35[2]	1064	-	1064	-	1064	-
1935-36	1627	-	1627	-	1627	35 (89)
1936-37[3]	3158	-	3158	-	3158	36 (91)
1947-48 (1/14-1/23)	1199	-	1199	871	2070	6 (15)
1948-49 (1/11-1/20)	1358	-	1358	130	1488	30 (76)
1949-50 (1/25-1/30)	533	312	844	90	934	48 (122)
1950-51 (2/3-2/16)	-	1434	1434	-	1434	13 (72)

[1] Ground counts done by patrolling selected drainages up to limit of observed elk sign. Aerial surveys were with a Piper PA-11 Super Cub.
[2] Nez Perce-Pete King-Deadman-Wild Horse survey of January 1935.
[3] December 1936 surveys, Nez Perce National Forest.

Winter conditions were considerably milder than in 1936, with 6 to 12 inches (0.15-0.30 m) of snow along the river. Elk were reported to be dispersed, with the largest herd being 130 head at Bald Mountain. A "conspicuous absence" of bulls was noted, thought to be attributable to either an actual shortage, premature antler loss making identification poor, or bulls occurring at extreme top limits of the range, with the occurrence of bulls at high elevations being the likely reason. A classification of sex and age was not made during the snowshoe survey. Elk seemed to be in good condition and browse use was considered light, except on Bald Mountain.

The January 11-20, 1949, survey was again a snowshoe count, by four men on the Lochsa District, following an advance scouting of the outer boundaries of the winter range by air. The Powell District also sent ground observers onto the winter ranges further east. Because the advance scouting was not done by people doing the ground count, "the accuracy and completeness of the final survey was seriously affected even to the point of reducing it to a pure guess in certain specific drainages." During the Lochsa count (surveys on the North Fork of the Clearwater were also in progress),

24- 30 inches (0.6-0.9 m) of powdery snow prevailed. The actual number of elk counted was higher than in 1948 by 300 head, but the total population was estimated to be lower.

Table 5-2. Classification of elk during surveys from 1934-1951, taken during census efforts reported in Table 5-3.

	Bulls	Cows	Calves	Unk.	Total	Comments
1934-35	87	308	139	107	641	
1935-36	272[3]	933	322	100	1627	severe winter
1936-37	101	183	118	250	652	
1948-49	229	457	123	549	1358	severe winter, starvation loss, 130 estimated not seen
1949-50	48	266	52	167	533	No spike bulls seen

[1] Ground counts done by patrolling selected drainages up to limit of observed elk sign. Aerial surveys were with a Piper PA-11 Super Cub.
[2] Nez Perce-Pete King-Deadman-Wild Horse January survey 1935. Nez Perce NF Game Surveys.
[3] Includes 76 spike bulls in 1935-36

The February 16-23 period brought rain which crusted the snow on the 24th, making conditions critical for game. Elk wintering in timbered areas were thought to be in poor condition and with less vitality than the elk using the shrub fields. A total of 60 dead elk were observed in the Lochsa during the survey, including 50 which apparently drowned and 10 which died of malnutrition. A supplemental wildlife condition survey report dated May 26, 1949, reported a total of 320 dead elk in the Lochsa, and the estimate of loss was 15-25% of the herd. The mouth of Sherman Creek was heavily used and 43 dead elk were located in its vicinity. Mortality on the North Fork Clearwater winter ranges was estimated at 49 head during the January surveys, lower than on the Lochsa. Substantial mortality of elk was apparent during this winter (Figure 5-1).

Figure 5-1. Dead elk that were found in or near the Lochsa River were hauled out and burned. Photo provided by Don McPherson.

Moore (1996) reported the surveys conducted during the 1948-49 winter. The crew from Powell reported 1120 elk while a crew from the lower range out of the Lochsa Ranger Station reported 1488. Subsequently, surveys in April reported that more than half of the elk had died in the lower Lochsa. Parsell (1990) reported that probably well over a thousand elk in the Selway-Lochsa country died that winter.

Weather records for the Fenn Ranger Station were used to develop a winter severity index for the 1948-1949 winter (Figure 5- 2). This index was developed using the maximum daily temperature subtracted from 32^0F. and adding the snow depth (Picton and Knight 1969). While snow depths at Fenn average 18 cm. (7 in.) in January and 13 cm. (5 in.) in February (Figure 2-2), depths reached 38 cm. (15 in.) in early December 1948, 64 cm (25 in.) in mid-December and over 89 cm. (35 in.) in mid-February. A winter severity index above 20 was maintained into mid-February. The mean severity index for Fenn over the 1948-2017 period of record was -12.7 in November, -1.8 in December, 3.1 in January, -15.9 in February and -18.4 in March. Records do not go back that far for Powell, but Moore (1996) reported 130 cm. (51 in.) of snow at Indian Grave Creek in 1948. Elk were pretty well confined to conifer cover with lower snow depths, or else to small areas where available forage was severely limited This was one of the most severe winters in the northern Rocky Mountain states in the century.

Table 5-3. Distributions of elk from 1935-36, 1947-50 surveys within the Lochsa River drainage within the Clearwater National Forest. The area covered did not include the Powell district.

Location	1934- 35		1935- 36 [1]		1936- 37		1947- 48 [2]		1948- 49 [3]		1949- 50 [4]	
	No.	%	No.	%	No	%	No.	%	No.	%	No	%
Fish Creek	33 [5]	5	57	4			210	32	132	10	139	16
Lochsa River			1236 [7]	76	391	39	235	33	703	52	500	57
Sherman Ck.			138 [8]	8			114	16	208	15	104	12
Boulder Ck.			196 [9]	12			147	21	315	23	135	15
Lower Lochsa	834 [6]	95	1356		604	61						
Totals	922		2983		995		706		1358		878	

[1] Page 3 of report of 14 May 1936, Winter game study-Clearwater National Forest, Orofino.

[2] Memorandum from E.F. Harry, Supervisor Clearwater National Forest of 6 Feb 1948 to wildlife management files, Orofino.

[3] Memorandum to Clearwater National Forest Supervisor from Lochsa Ranger L.F. Hartig dated 11 February 1949, Pierce, Idaho.

[4] Memorandum from L.F. Hartig to Forest Supervisor of 6 February 1950, includes actually observed and estimated missed, Pierce Idaho.

[5] Taken from Coster and Parsell 1946. Deadman to Bimerick Creek surveys of 22 January-10 April, 1936.

[6] Landall and Fager (1935:48) excludes 142 seen in Middle Fork. From Boulder Creek to confluence of Lochsa and Selway Rivers.

[7] Includes Grouse Ridge.

[8] Includes Lone Knob, Lochsa Point.

[9] Includes Noseeum Creek, Bald Mountain Creek, Holly Creek, Beaver Flats, and Macaroni Creek.

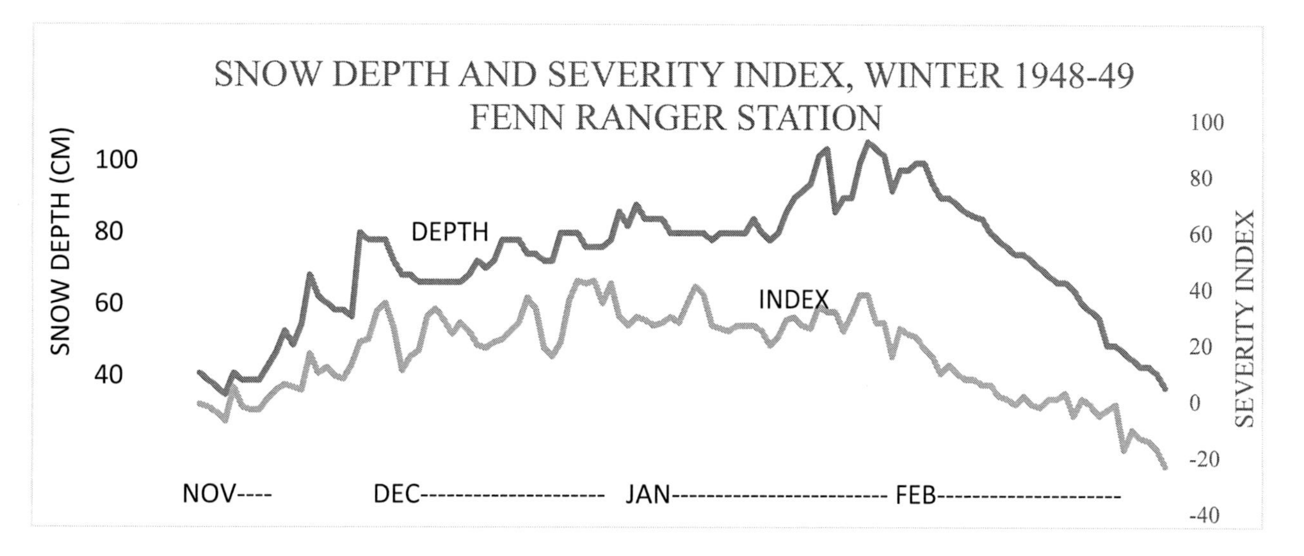

Figure 5-2. Fenn Ranger Station snow depth and winter severity index 1948-1949 winter. The mean severity index for the 1939-2016 period was -12.7 in November,-1.8 in December, 3.1 in January,-15.9 in February and -18.4 in March.

First mention of use of a helicopter to survey particularly inaccessible areas was made in the 1949 report. Additionally, a study of browse conditions in the Lochsa was recommended to determine the "urgency of recommending the abandonment of the Game Preserve" (Weyerman report, Clearwater National Forest, dated March 23, 1949).

An April 7, 1949, press release concerning the wildlife survey reported low losses of elk due to predation and predators at the time of the survey, although greater losses were expected later. A May 18, 1949, memo to the Forest Supervisor from the Lochsa District Ranger reported the April 11-18 survey which documented substantially more mortality and browse damage, and that no calves were observed among the 155 animals that were seen alive during the survey. Space (1964) reported that the Clearwater Forest personnel generally believed that the elk population reached its peak in 1948, was substantially reduced by the severe 1948-49 winter, and that hunting pressure had kept the population below earlier levels up through the 1950s. A memo to Director T. B. Murray, IDFG, dated March 31, 1949, compared the 1935-36 survey with the 1948-49 survey which "indicated the trend of elk to increase and deer to decrease during the past decade on the Clearwater drainage surveyed."

Evidence that Forest Service personnel were beginning to consider action to increase winter forage was provided in a May 25, 1949, memo from E.F. Barry, Supervisor, Clearwater National Forest, to the Regional Forester in Missoula. While reporting that the Lochsa was "overstocked with elk for snow conditions such as prevailed last winter," experiments in cutting browse by Lochsa District Ranger Hartig to reduce the height of the browse and stimulate resprouting were ongoing. The Supervisor wondered in the memo if this wasn't important enough to deserve an expenditure of Pitman-Robertson Funds, which are the Federal Aid to Wildlife Restoration funds used by IDFG to manage elk and other wildlife populations. As will be seen, cooperative efforts by the two agencies resulted in substantial winter range improvement activity by using prescribed fire, almost twenty years later.

IDFG and USFS personnel in the late 1940's were obviously interested in using airplanes for elk census, but considerable controversy occurred. Initially planes were used to supply ground crews in the North Fork of the Clearwater where distances to the elk ranges from stations occupied in winter were great. Concerns over the inability to locate elk in timber from the air were apparent, and safety was always a factor. A comparison of a 1950 ground count with an aerial survey on 6 February, showed 242 elk from the air and 445 from the ground. Still, experiments were made with a Piper PA-11 Super Cub and a Cessna 170 "at no expense to the Forest

Service, and it was clearly understood by ourselves and our dependents that the project was entirely on our own and in event of accident the Forest Service was in no way responsible. The afore is stated because the plane we used did not comply with the Region 1 supplement of the Safety Code and did not consist of going from one landing field to the other" (Forest Service memorandum on file, Clearwater National Forest, spring 1950). The Super Cub, with its lower cruising speed and greater maneuverability, tandem positioning of pilot and passenger, was considered to be the "proper ship for the work." The cost of surveying the North Fork of the Clearwater and the Lochsa was estimated to not exceed $250.00 which may be compared with the $12,000 estimated to survey the Lochsa with helicopter in recent times (Unsworth and Kuck 1991). Cost of the Super Cub in 1950 was $12.00 per hour.

Subsequently, V. L. Erickson, District Ranger at Bungalow on the North Fork, concluded that:

> "an air count was more accurate than a ground count if properly made. The ground count might see more elk in timber and bottoms but would miss more on the open slopes. Additionally, sex and age of animals could be obtained if a mountain pilot and trained observer were employed."

The 25-31 January 1950 count, however, was a ground count. Snow at the Lochsa station was crusted at 32 inches (81 cm) and was 49 inches (122 cm) total depth. A16 inch (41 cm) snowfall on 25 January covered the earlier crusted snow and made snowshoeing difficult. The estimated total of elk was over 500 animals lower than the previous year's estimate, reflecting the poor conditions for counting from the ground. However, not a single spike bull was observed, and the observed cow:calf ratio was 100:19. The conclusion was that the Lochsa District had only half the elk that were there in 1948. Additionally, the road up the river, which extended in 1950 to Beaver Flat, was recommended to be closed at Pete King Creek to prevent disturbance to wintering elk by sightseers and poachers.

Evidence that there was disagreement over whether there was sufficient winter range among the land and game managers was provided in a February 6, 1950, memorandum from the Lochsa District Ranger to the Supervisor of the Clearwater Forest: "Regardless of the opinions of some of our game managers to the contrary, the winter feed situation on the Lochsa River is desperate. To attempt to keep this fact from the public is fatal." Recommendations included "...game managers make use of field reports in arriving at conclusions in regard to game conditions." Undoubtedly Forest Service personnel, trained and conditioned to examining habitats and primarily responsible for their condition, would be alarmed at the observed heavy browse use. Population managers concerned with harvest and numbers might be expected to focus on hunter harvest and other mortality as the major problem and minimize the consequences to the population of the occasional severe winter. However, jointly written memoranda and news releases provide evidence that efforts to coordinate activities and present a common front to the public were in play. It may be noted again that while the earlier surveys in the 1930s were carried out only by Clearwater Forest personnel (a state game warden participated on the Nez Perce National Forest surveys starting in 1935), the post-World War II censuses were joint ventures between IDFG and USFS, with a federal predator control man from the Biological Survey often included as well.

The February 1951 census is notable primarily for its being the first one done completely by the air in the Lochsa. The area surveyed included the entire drainage to a few miles above Warm Springs Creek. Snow depths were 13 inches (33 cm) at the Lochsa Station, so the winter was considerably less severe than the previous one. Flights were done on 180-300 m elevational contours along 13-16 km segments of the drainage. However, a Curtis air sedan was employed for most of the flights and a Stinson was used for the last two when the Air Sedan developed engine trouble. One of the recommendations was to ground census elk in green timber, and several drainages known to contain elk were not considered adequately surveyed. IDFG authorized 20 hours of flight time of which 18-20 hours were used, and personnel from both agencies were involved.

These early surveys provide written records of the prevailing opinions and information on the Lochsa elk population and its habitat. They represent the initial efforts to formally count the elk population and examine its habitat. As early as 1936, malnutrition loss and concern over inadequate forage conditions for elk were reported. While in 1935 and 1936, the Pete King fire of 1934 could be blamed for the poor forage conditions, by 1947, growth of shrubs beyond the reach of elk, and scarce forage in areas of "green timber" where elk wintered became the concern. The prevailing opinion that elk reached a high in the drainage in 1948 and that the severe winter of 1948-49 precipitated a major die-off is reminiscent of many other big game populations of the late 1940's and 1950's which apparently reached highs and declined, often with dramatic winter die-offs involved.

However, the age-old controversy over the cause of the increase and decline was again raised. Portions of the Lochsa River were included in the old Selway Game Preserve and elk were not hunted until 1935, and then by permit obtained through a lottery drawing which restricted the harvest. At the same time predator control using 1080 and strychnine was ongoing, primarily to reduce coyote predation on a dwindling deer population. While

Forest Service personnel were recommending increased harvests of elk in the Lochsa at this time, many reports alluded to the scarcity of deer and a need to reduce or eliminate hunting for them. For elk, the question of whether reduced human and predator pressure was responsible for the increase or whether the increased winter forage base from 1936-1948 was primarily responsible remains. Reduced hunting pressure during the war years, emphasized by restrictions on hunting in the Preserve and a general inaccessibility of the Lochsa elk prior to completion of the highway, may have all interacted to allow the population to take advantage of the increased forage resource and to increase.

The quality of information obtained from these surveys would of course be suspect in view of current census methods and knowledge. There was controversy over their value even while they were being used. There was no way to evaluate the accuracy of the counts and estimates of animals not seen were not reliable. While personnel were obviously aware of the biases and problems, occasionally numbers obtained were taken as absolute. In all probability, estimates of missed animals were lower than actually occurred.

However, their value in providing assessment of distributions, sex and age ratios, and habitat conditions should not be underestimated for their influence on subsequent activities. The census estimates themselves may well track population trends, even if their accuracy and precision is suspect. The surveys identified problems and suggested solutions, thus providing a basis for what subsequently transpired.

In the 1950s, emphasis in being involved in the information gathering process gradually shifted to IDFG from the USFS. The Clearwater Game and Range Study conducted by Elmer R. Norberg and Lester Trout from October 1, 1954 to September 30, 1957, stands out as the definitive investigation of the decade. The primary impetus for this study was the proposed building of two high dams in the Clearwater River drainage by the U.S. Army Corps of Engineers. Bruces Eddy on the North Fork of the Clearwater at Ahsaka eventually was the selected site for the Dworshak Dam which inundated approximately 4,352 ha distributed along 78 river kilometers of critical winter range for elk. If the Penny Cliffs site on the Middle Fork of the Clearwater had been selected, approximately 24,576 ha of winter range in the Middle Fork, Lochsa, and Selway River drainages would have been eliminated. This dam, plus the yet-to-be completed Lewis and Clark highway through the Lochsa, would have undoubtedly had major effects on big game.

Needless to say, Norberg and Trout (1957) recommended that the dams not be built. Needless to say, that recommendation was not followed, but perhaps the decision to dam the lower North Fork was influenced by the ramifications of the alternative of inundating both the lower Selway and Lochsa, plus a portion of the Middle Fork.

However, Norberg and Trout provided additional recommendations concerning mitigation of the loss by acquisition of adjacent lands and funds to study management problems arising from the dam, which were implemented. These recommendations resulted in habitat management programs along the North Fork (Meske 1972) and have included the acquisition of the Howard properties along the Snake River on Craig Mountain south of Lewiston, some 40 years later.

The Clearwater Game and Range Study included the first comprehensive and systematic attempt to census elk in the Clearwater River drainage with a helicopter. As happened in 1950 when ground counts were compared with fixed-wing surveys, a comparison between the fixed wing and a helicopter was made in March 1955. A decision to use the helicopter for the main census and to use fixed-wing craft "only on selected trend count areas" was made. The decision required that fuel be stored at selected landing sites in the census area, because the helicopter required about 15 gallons of fuel per hour and carried only 30 gallons (a 47D-1 Bell helicopter was used). A total of 1,800 gallons of fuel was placed in the field and an additional 572 gallons was brought in and used during the flights.

The census planning was thorough, and involved not only fuel placement prior to its execution, but evaluation of the validity of the count by making pre-census reconnaissance flights and ground surveys in the lower North Fork. The old method of snow-shoeing side drainages in the sampled area and use of a snow tractor along the main travel routes was used. Two winters, 1955-56 and 1956-57 were required to census the entire Clearwater designated for investigation, using two helicopters. The second helicopter was used for safety reasons as well as to cut census time. The possibility of forced landings in remote areas was real, census crews were often stationed in remote areas where contact with others was irregular.

The caching of fuel in remote areas when helicopter census of big game is to be made, and use of remote, isolated facilities in the Idaho back country remains in practice today. Now, safety precautions require radio checks at 60-minute intervals with permanent stations, to advise of location and status of flight. Aerial game census remains a notoriously risky activity despite use of the best equipment and safety practices possible, and hardly a year goes by but what an accident resulting in lost personnel and equipment occurs somewhere in the west and north. But in 1954, the decision to use helicopter equipment in a remote inaccessible region required extensive consideration of the risk, balanced against an obviously compelling desire to do a thorough job. The project established a pattern which has lasted essentially for 30 years until improved census techniques again were developed, using the Lochsa as the location once more for the testing and the initial work.

The census included the entire Lochsa drainage plus the Middle Fork to Kooskia. A total of 4,321 elk were included in the Lochsa count, of which 90 were observed on the Middle Fork down to Kooskia. The count was made in 1956-57, a relatively mild winter when animals were observed to 1,524 m elevation in mid-January. After a thaw and freeze, game was concentrated at or below 1,219 m, and the census was started shortly after February 7. Snow depths were high (127 cm at Moose Creek Ranches on the Selway on February 7), but weather turned mild and rainy which resulted in baring of many slopes and some upward movement of elk resulted. Sex and age of elk observed were not determined during the survey because the biologists did not think they could accurately classify them. Perhaps this was because they did not fly low enough.

The 4,321elk seen in the Lochsa unit in 1957 may be roughly compared with the 5,112 elk observed in 1949, and the 3,158 figure for 1937 in the entire drainage, suggesting an increasing population in the late 1930s and early 1940s and a decline in the 1950s (Numbers in Table 5-4 include those seen on Clearwater National Forest). This was the conclusion of Norberg and Trout (1957). It must be recognized that comparisons between years using different census methods is highly risky. The 1948-49 winter was severe and conditions for the ground surveys to observe elk may have been excellent. Conditions for the 1957 survey may not have been as good. Unsworth et al. (1993), using current census technique which incorporates corrections for visibility bias among sexes relative to group size and vegetation cover, estimated that approximately 56% of the elk are observed when helicopter census is employed in the Lochsa. If this is the case, and assumptions used in deriving that estimate can be directly applied to the 1957 census, then 5,781 elk may be estimated to have been present in the Lochsa during that year. Norberg and Trout (1957) estimated that they observed 80% of the population, which would make the total 5,401. Fall pre-hunting season estimate was 8,000 animals. Estimates for the 1985-89 period of between 2,000 and 3,000 elk observed and 4,000 to 4,500 elk present suggest a decline from the 1957 period. Norberg and Trout (1957) concluded that the elk were at as high levels as forage supplies would support at that time. We conclude that an increase from the turn of the century and before lasting into the 1950s and a general decline afterward represents the long-term trend for the Lochsa elk population. An increase after 1975 when hunting of cows and calves was eliminated will be placed in the perspective of this general trend later. The declines occurred during the 1996-97 winter when IDFG estimated a +50% loss of the population. This decline coincided with the introduction of wolves. Predation is what has kept the populations from rebounding after the winter losses.

Table 5-4. Distribution and numbers of elk observed in February 1957 from a helicopter in the Lochsa, after Norberg and Trout (1957)[1].

Area	South No.	North No.	Total No.	Per Cent of Total
Middle Fork below Lowell	90	-	90	2.1
Lowell to Deadman Ck.	552	219	771	17.8
Deadman to Otterslide Ck.	173	698	871	20.2
Otterslide to Boulder Ck.	590	45	635	14.7
Boulder to Lake Ck.	922	445	1367	31.6
Lake Ck.	-	247	247	5.7
Lake Ck. to Powell	211	101	312	7.2
Above Powell	28	-	28	0.6
Totals	2566	1755	4321	99.9

[1] Totals presented in Norberg and Trout (1957:101) are corrected above.

Salt

Controversy over the value of salt in benefitting big game in Idaho, and the Clearwater River drainage in particular, has been ongoing since the 1930s. The practice of placing salt out on winter and spring ranges has roots in animal husbandry and extends into wildlife management in several aspects. The 1934-35 Nez Perce winter game study (Landall and Fager 1935) states the issue:

> "Observing activities of elk and deer about the several natural and artificial salt licks has led to a confirmation of the belief that proper salting will help to solve the distribution problem and well as to supply amply the salt requirements of the game."

Wherever salt licks, natural or artificial occur, they receive heavy use by big game, particularly in spring (Figure 5-3).

The "distribution problem" has to do with the heavy utilization of winter forage on restricted, exposed slopes, and the assumption that game would be encouraged to move away from such sites by proper salt placement. Concern over salt placement was raised in the 1935 report because much of it dissolved prior to use. The 1936 report (Coster and Parsell 1936) recommended higher quantities of salt for both summer and winter ranges and criticized Idaho Fish & Game for distributing too little. This report recommended that all licks and hot springs should be fenced against entry by big game to reduce infection with diseases and parasites, contrasting with the recommendation to apply more salt. A total of 88,200 pounds of salt was considered required by big game in the Lochsa-Selway complex. The 1937 report (Fry and Parsell 1937) recommended 2 tons of salt be placed on the spring game ranges.

Young and Robinette (1939) investigated game use of the Wounded Doe licks in the Selway River drainage. They described the area as being denuded of nearly all plants except trees over 6 inches in diameter in a 15-acre area. Elk trails that converged at the licks could be followed for 3 to 4 miles away. Purposes for applying salt included drawing game away from winter ranges earlier in spring, fulfilling the mineral requirements to maintain health and vigor of big game, and to obtain better distribution of elk on summer range. Their recommendations were that salt be distributed.

Moore (1996) describes the program to lure elk off winter range using salt. While use of pack animals was tried, on March 27, 1948, salt was dropped at higher elevations that were inaccessible in early spring by airplane. Drop spots were typically on ridges or a basin heads or in forest openings. About 8,000 pounds of salt could be dropped, using three trips in a Ford Trimotor. The salting program eventually included much of the back country in Idaho. The program continued until 1960, when Kindell (1958) and Dalke et al. (1965) concluded that salt grounds caused elk to redistribute to their proximity but did not encourage earlier or more rapid movements off winter range when compared with areas without salt. Today, some individuals claim that it enhances calf growth and survival and reduces parasitism, but no substantial evidence was available. While stockmen commonly use salt to distribute livestock, and wild game will use the salt, especially in spring, the program was abandoned as expensive and not working as planned.

Figure 5-3. Elk trails going down to the Weir Creek hot springs, a natural lick, on the right of this photograph. U.S. Forest Service 1960.

CHAPTER SIX

INVESTIGATIONS OF PREDATION

Predators have always been a source of concern in the Clearwater region, and the Lochsa in particular. The annual game report of the Nez Perce National Forest of 1922 concluded that "the greatest single factor in the protection of game animals at this time is the destruction of predatory animals." The record of surveys and reports that started in the mid-1930s shows how changes in attitudes towards predators has affected their management. It also shows that concerns over heavy browsing were reported in the 1930s. The 1936 Forest Service survey of winter severity recorded numbers of coyotes and cougars (one wolf was estimated to be present in the Musselshell District based on sign). Poisoning and shooting coyotes was ongoing, and a recommendation to request a predator contact person from the Biological Survey to poison coyotes in the Lochsa was made, even as the "Lochsa River drainage this winter was greatly overstocked in accordance with the feed present" (Eddy and Rauma 1936:11).

The 1934-35 Nez Perce Report concluded that coyotes were not present in numbers that unduly affected the game, but that they were not yet under control. An eradication program for another two years was recommended. No cougars were observed in the Lochsa but were apparently present in low numbers. This report provided extensive discussion on cougar:

> "Though one cougar makes an apparently enormous kill each year, the fact remains that of all the deer or elk taken, the largest percent are those animals rendered unfit by disease or injury. Shaver records additional observations showing the cougar to be a naturally 'lazy' individual. He will make several short runs at game until he finally encounters an unwary one or handicapped animal and makes a kill."

The 1934-35 report also states that it would:

> "not seem scientifically justifiable to advocate extinction of cougar or coyote. The cougar, more than the coyote, is important from the standpoint of keeping game herds free from diseased or weak stock. Cougars continue to make kills on big game during the warmer seasons. Of course, the number

> of cougar must be very small owing to their killing habits and if at any time game kills become excessive in certain local areas a hunter should be sent immediately to hunt down a cougar or two."

An estimate of three coyotes killed by the game crew and 50 poisoned was provided in the report. The report stated:

> "Unless a worthwhile kill can be made, the Forest Service can expect to get no returns for the food stuffs donated as a cooperative share to the Biological Survey."

Obviously, the number of predators killed was an incentive for both the U.S. Department of Interior Biological Survey and the Forest Service. The Biological Survey was responsible for predator control. Much of the concern over predators was related to scarce deer rather than the elk.

In 1935-36, four members of the Biological Survey and one hunter working for the Fish and Game Department were assigned to the Nez Perce Forest. The state hunter was assigned to "exterminate" cougars. Additionally, a "large quantity of poison was provided for the use of our sturdier men in the field." (Coster and Parsell. 1936). However, poisoning was considered "decidedly unsatisfactory" because activities of field personnel precluded checking bait stations, so the results were unknown. Thus, the recommendation that predator control should be turned over to the Biological Survey was made.

The 1936-1937 surveys (Fry and Parsell 1937) indicated that coyotes were abundant while bobcats were scarce, and cougars rare. The report noted that poisoning coyotes was controversial among the commercial trappers and the agencies that were killing the predators:

> "Criticism of the estimate is continually aroused by commercial trappers, who, in fact have no concrete facts to present showing that poisoning hinders their business, other than the killing of coyotes."

Evidently the trappers and others were aware that bait stations were visited by other species than coyotes in the region, probably because other species were seen dead within the vicinity of the bait stations. Strychnine pills were distributed in the Lochsa drainage in the 1949 winter by Forest Service personnel. The predator control person did go along on surveys of the 1080 stations in the 1940s and early 1950s. Both cougars and coyotes were scarce based on the 1950 survey.

Grizzly bears were nearly extinct in the early 1900s in Idaho and the last bear killed in the Clearwater was in 1926 on Wallow Mountain according to Space (1964) and Moore (1996). Space felt that trappers had virtually extirpated grizzlies in the Clearwater by about 1900. The 1934-35 report (Landall and Fager 1935) stated that grizzly bears were so few and far between that an absolute closed season was recommended to perpetuate the animal.

Predators continued to be considered a major reason for the low calf survival in the Lochsa and the rest of the upper Clearwater drainage. While the vast shrub fields following fires were either changing to coniferous forest or had lost value as forage, habitats were also recognized as contributing, poor calf survival prompted evaluation of physiological condition of penned elk, road-killed elk, illegal hunter-kills, and wild elk captured in the Lochsa (Vaughn et al. 1973). The penned elk included one 6 year-old pregnant cow and two bull calves. They were fed high quality alfalfa pellets supplemented with milled grains and oil meals.

In winter 1969-1970, 37 adult female elk were all examined for pregnancy using rectal palpation, similar to pregnancy testing for cattle. All but one elk was pregnant, indicating a high pregnancy rate. No evidence of aborted fetuses, weak, or sick calves was found in the study areas. Blood serum collections yielded comparisons of blood chemistry between one wild 7 year-old pregnant female and the penned elk. Blood urea nitrogen levels, indicative of the quality of the diet, were higher in the captive elk than in the wild elk, but no indications of nutritional imbalances suggesting that poor nutrition was a cause of the low calf survival. Diseases that would cause abortion such as brucellosis were also eliminated. Vaughn et al. (1973) reported that predation was the likely cause of poor calf survival after concluding that they had eliminated the nutrition and disease as limitations. The abundant predators including cougar, bear, coyote, and bobcat were thought to be advantaged in the pursuit of young elk calves in this steep and heavily vegetated terrain.

The Coolwater Ridge Investigations

Idaho Fish & Game initiated further investigation of causes of calf mortality in 1973 in the Coolwater Ridge area on the lower Selway and Lochsa (Schlegel 1976,1986). This work was the first effort to assess calf mortality using radio telemetry and the procedure for attaching radiocollars to calves so the collars expand as the calf grows has been used in many studies of this kind since. Objectives of this work were to determine condition of newborn calves and causes of mortality. Calves were captured using a helicopter, fitted with radio-collars that contained a motion unit that changed from an irregular signal to a continuous signal if no movement occurred during a four hour period. Radio-collars were modified to fit calves with a neoprene impregnated nylon strip sown onto the collar to allow it to stretch as the calf grew (Figure 6-1). When a continuous signal occurred, the transmitter was located through a ground search and the carcass was examined and cause of death was determined. Puncture wounds, claw marks, subcutaneous hemorrhage, hair, tracks, and feces were recorded and usually provided evidence of what killed the calf. Complete necropsies of the carcass done at the Veterinary Pathology program at Washington State University were performed when field examinations were inconclusive.

A total of 67 calves were captured over the 1973-1975 period of study. Nine calves lost their collars, 34 calves were killed by predators, 5 calves that were captured were subsequently rejected by their dam, and 1 calf was killed by hunters. Black bears accounted for 25 calf mortalities, cougars took 5 calves, and 4 were taken by predators that could not be identified. Most of the predation occurred prior to June 14 of each year (Figure 6-2).

Schlegel (1986) compared calf weights from the Coolwater Ridge area with calf weights taken in the Gallatin River drainage by Johnson (1951). Calves in the Idaho area weighed an average of 37 pounds at one day or less of age, while the Gallatin River weights averaged 32.5 pounds. At eight days, the Idaho calves weighed on average 77.3 pounds and the Gallatin calves weighed 53.2 pounds. No evidence of nutritional deficiencies in the

Figure 6-1. An elk calf with the collar that has an expansion portion that will expand and eventually break as the calf grows. Schlegel photograph.

in the Idaho calves was apparent. During that period, the Gallatin population was at high density following the severe winter of 1948-1949, both of which would be related to the lower calf weights. The work by Thorne et al (1976) verified that the Coolwater Ridge calves were within a normal weight range for healthy calves.

An experiment to determine whether removal of black bears in the Coolwater Ridge area would result in increased calf survival was initiated in 1976 with the removal of 75 bears (Schlegel 1986). Two winters following the removals increased winter cow:calf ratios to 61 in 1977 and 51 in 1978 (Figure 6-3). Movements of 12 bears following their capture and transplant to other areas in Idaho were followed (Figure 6-4). Five bears returned to the Coolwater Ridge study are, including three that were transplanted south of the Salmon River and the Sawtooth Mountains.

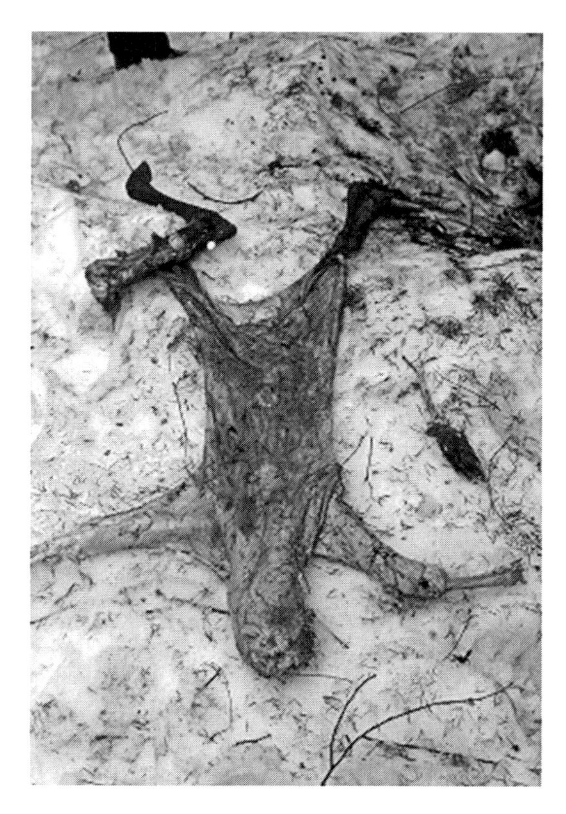

Figure 6-2. Bear predation on elk calves often results in the skin being left as the bear eats the rest of the calf. Schlegel photograph.

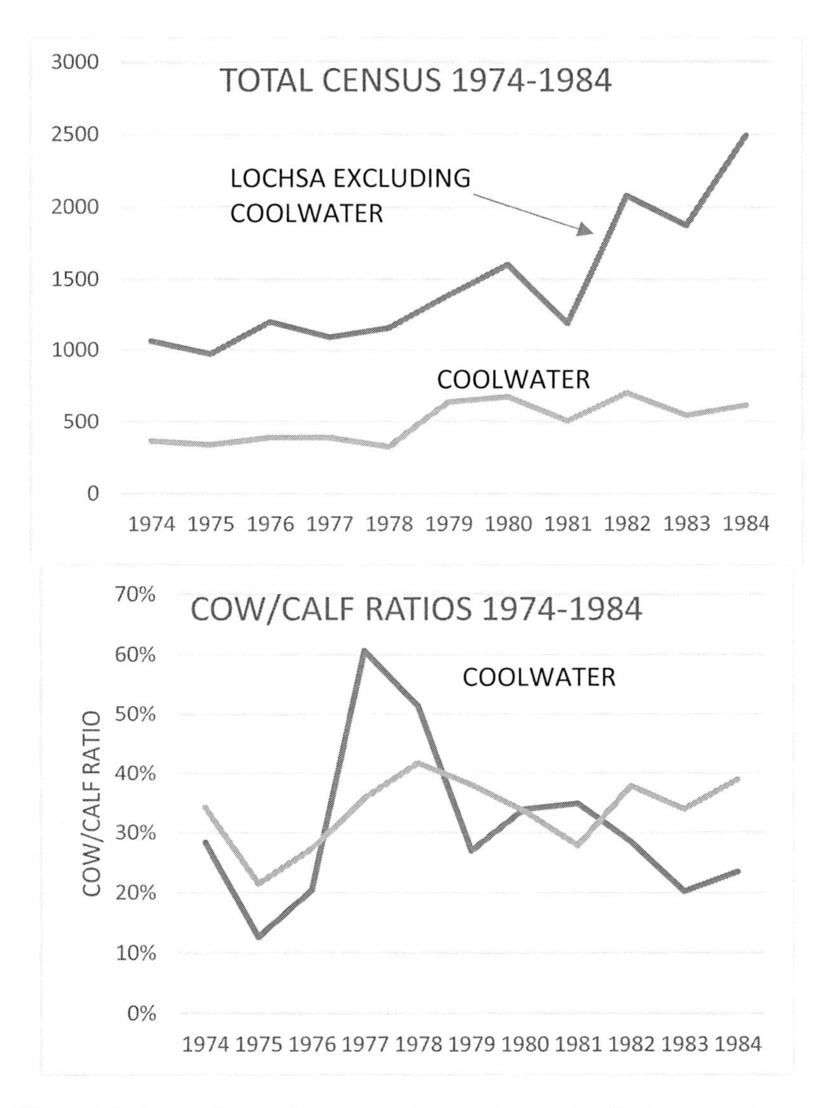

Figure 6-3. Comparisons of census and age ratios on the Coolwater study area compared with the rest of the Lochsa, 1974-1984 (Schlegel 1986).

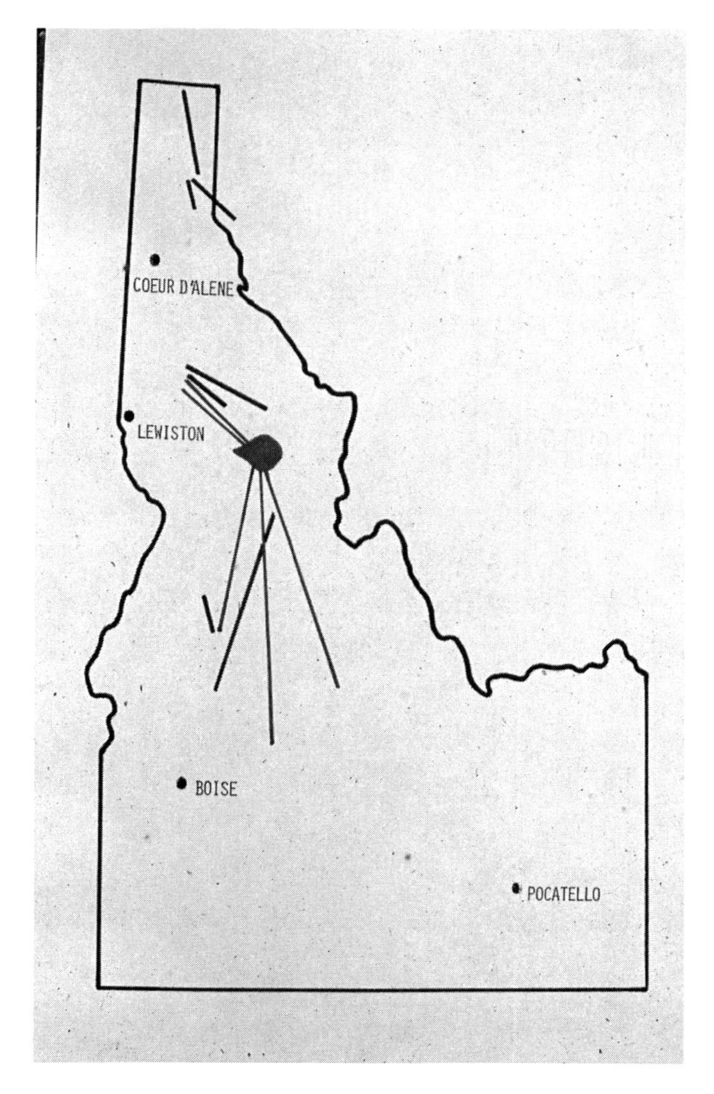

Figure 6-4. Directions of 12 black bears removed from the Coolwater Ridge area to other parts of Idaho. Red lines indicate bears that returned to the Ridge and black lines indicate bears that remained in the area where they were placed.

Idaho Fish & Game continued work from 1986 through 1991 with 121 elk including 101 bulls and 20 cows captured, ear-tagged, and radio collared (Unsworth et al. 1993). This work was focused on adult elk rather than calves. Hunting seasons included an either-sex archery hunt in September and a 26-day rifle season in October with cow take limited via controlled hunt. Survival rates ranged between 0.552 to 0.733 for bulls and 0.778-1.00 for cows during the July 1987-June 1991 period. As expected, rifle hunters accounted for the majority of the mortality which included 41 bulls, 8 wounded and died. Two radio-collared bulls were taken by archers, and 4 bulls were wounded and died. Poachers took 3 bulls and 6 were classified as other mortalities. Surviving bulls used areas with lower open roads and hunter densities, in steeper, more broken terrain, and therefore less accessible to hunters. Cow mortality included 2 rifle kills and 3 other mortalities for a survival rate of 0.89. This work indicated that predation on adults in the Lochsa was not important in those years.

Hunting Mortality

Hunter harvest accounted for over 90% of the mortality of adult bulls excluding yearlings (Table 6-1). The remaining causes included vehicle collisions, diseases, and accidents. Survival rates for adult bulls were 60% per year. These data were for the years 1986-1995 and included 400 individuals.

Table 6-1. Survival rates and causes of mortality in elk Lochsa and North Fork Clearwater, ummarized from Unsworth et al. (1993) and Zager et al (2006).

	No.	Survival Rates and Causes of Death							
		Rate	No.	Harvest[1]	Other[2]	Bear	Cougar	Wolf	Total[3] predation
ADULT BULLS									
1986-1990	169	0.6		90.7	9.3	0	0	0	0
1991-1995	231	0.63		98.8	10.2	0	0	0	0
ADULT COWS									
1986-1990	46	0.89	5	40	60	0	0	0	0
2005-2006	44	0.80	25		4		4	80	96
CALVES									
1997-2001,2004	97	0.26	101	0	10	43	42	2	91

[1] includes 11 wounded, 7 during rifle season and 4 during archery season.
[2] Vehicle collision, disease, accidents, unknown.
[3] Includes unknown predator.

Further Investigations into Mortality Causes

Idaho Department of Fish and Game continued the studies with a comparison of survival rates and causes of mortality in two separate study areas (Zager et al. 2006, White et al. 2010). The project was presented to the public earlier so hunters would be informed of the changed regulations, resulting in an unanticipated increase in harvest on the control area two years before the initiation of the study. The South Fork Clearwater area elk showed higher calf survival, relatively stable elk population size, and higher harvests of bears and cougars than the Lochsa. This study lasted from 1997-2004. On the Lochsa treatment area, black bear and cougar harvests were increased and calf survival was assessed starting in 1999. Survival estimates for calves were 19% in 1997, 28% in 1998, and 32% in 1999, 26% in 2000 and 57% in 2001. The treatment area showed survival rates of 33% in 2000, 50% in 2001, and 70% in 2004. Estimates on the South Fork study area were the opposite, with reductions in predators showing reductions in calf survival. An analysis of predictors of calf survival showed the black bear harvest, estimated mass at birth, presence of a shrub field, and the interaction of shrub field with time since birth were the best predictors of ten variables examined.

White et al. (2010) and Zager et al. (2006) concluded that predation by black bears was the most important proximate mortality factor for elk calves from birth through August in north-central Idaho (Table 6-1). Because that mortality was additive, aggressive black bear management improved summer calf survival. Zager and Beecham (2004) reported that black bear predation can limit the size of ungulate populations but will not regulate them. Bear predation may be additive at low ungulate population densities and then can become compensatory with other mortality factors at higher densities. Black bear harvest had a stronger direct effect on summer calf survival than did cougar harvest, although both effects were detectable. Because calves are generally most susceptible to bear predation in the first 28 days of life, spring bear hunts that reduce bear densities or at least disrupt bear activities may be most effective if directed at the individuals that are the predators. Increasing autumn bear harvest may have a similar, though less immediate and predictable, effect. Reducing black bear harvest may decrease calf survival, and this strategy could be used where the goal is to slow population growth or decrease elk population size. Calf birth mass and habitat structure also influenced calf survival in this study. Thus, addressing depressed elk recruitment with predator management alone may not be effective in achieving calf recruitment objectives. Improving elk habitat may increase physical condition in adult female elk, which should lead to heavier calves presumably the most vigorous, with higher summer survival and, perhaps, winter survival. Based on these results, structural characteristics of good-quality elk habitat may also contribute to calf survival by improving escapement.

Observed differences in survival of calves on the two study areas also shows that different habitat conditions can affect management objectives, and treatments to manipulate elk populations and should be monitored to verify that objectives are being met. Weather conditions, both during the hunting season and at other critical times including late summer and late winter/early spring will also affect management.

Adult cow survival rates were 0.89 for the 1986-1990 period and 0.80 for the 2005-2006 years. In 2005-2006, 80% of the mortality was attributed to the wolf and 4% to cougars. No hunting of cows occurred during these time periods. Bears and cougars accounted for 43 and 42%, respectively, of the calf mortality while wolves accounted for 2%. Survival rates for calves during the 1997-2001 and 2004 were 26%. Not all mortality causes where predation was implicated could be identified to a specific predator and were classified as unknown. However, it was obvious that bear predation was important on calves, cougars were important predators on calves and adult cows, and wolves were primarily taking adult cows and older calves. Zager and Beecham (2006) concluded that bear predation appeared to be additive when ungulate populations were at low densities and became compensatory with other mortality factors when ungulate densities were high and approaching carrying capacity based on forage.

In the Lochsa, bear predation appeared to be limiting the population along with other predators. It appears that cow elk and black bear seek the same phenological zone in spring. On a late winter late spring, the zone is limited to a lower elevational range, making the chance encounter of a bear with a calf higher. During lighter winters and earlier springs, the range of this phenological zone is larger and the chances of an encounter are not as great. However, some bears were known to key in on calving areas in these investigations.

Gaillard et al. (2000), in a survey of dynamics of large ungulates including elk and red deer, concluded that predation is usually the main source of early mortalities with pre-weaning survival being less than 50% and can be as low as 1%. Survival of neonatal elk to 3 months declined following hotter previous summers and increased with higher May precipitation, especially where wolves and/or grizzly bears occurred (Griffin et al. 2011). Bear predation was the dominant cause of mortality in this age class in studies of 12 populations. Wolf predation was low for neonatal calves less than 3 months old and could be a mortality source that was compensatory with other sources of mortality, rather than adding to the total mortality. The composition of the predator community, whether it included grizzly bears and wolves or not, and when coyotes, cougars and black bears were the only predators, was considered important in the survival of elk calves (Griffin et al. 2011). In the Lochsa, the earlier investigations were at a time when wolves were absent, but calf mortality, primarily attributable to the black bear, was still high.

Black Bear

Idaho Fish and Game started requiring bear and cougar hunters to register their take in 1982, so a 36-year record has accumulated (Beecham and Rohlman1984, White et al 2010, Hayden 2018). Two-bear seasons and $25 nonresident tags became available in 1998, more than doubling the take for the 14-year period ending in 2012. Lochsa bear harvests for that period averaged 138 with the minimum take 102 and the maximum 166. There is no trend in the data over the period, and hunting conditions and berry crops, probably affected the differences in take between years (Figure 6-5).

The data sets are also developed for Data-Analysis Units that have similar characteristics and allow for more comprehensive analysis, since the traditional big game units (Unit 12 is the Lochsa) don't have enough harvest to allow for analyses. The Lochsa and North Fork Clearwater DAU 2A provides harvest records that allow for a look at the overall trend, the trend by sex, age structure, and median age of the harvest (Figure 6-5). One of the assumptions in using this information is that since the harvest is not restricted with quotas or permits, it likely reflects population trends, although how closely would require extensive work with captured animals over enough years to find out how accurate this assumption is. Direct observations of predator numbers requires use of radio collars, aerial observations, and is expensive. It appears that the two drainages supply approximately equal proportions of the harvest.

Number of bears harvested in this DAU for the 2000-2012 years show no trends. The mean take per year was 275 bears with a range of 237-334. Males averaged 182 per year and females averaged 92 per year, showing the emphasis on taking males and also that fewer females were available because sows with cubs were illegal to take. Spring hunting took an average of 240 bears, with males again being most heavily harvested, averaging 159 per season. Females were a bit higher in the spring take, averaging 81 per season. Much of the spring hunting was with bait, while the fall season would include incidental take by hunters after elk or deer. Hunters are required to have a permit, put the bait in a 55-gallon metal container or smaller, or an excavated hole and remove all materials within a week after the season closes.

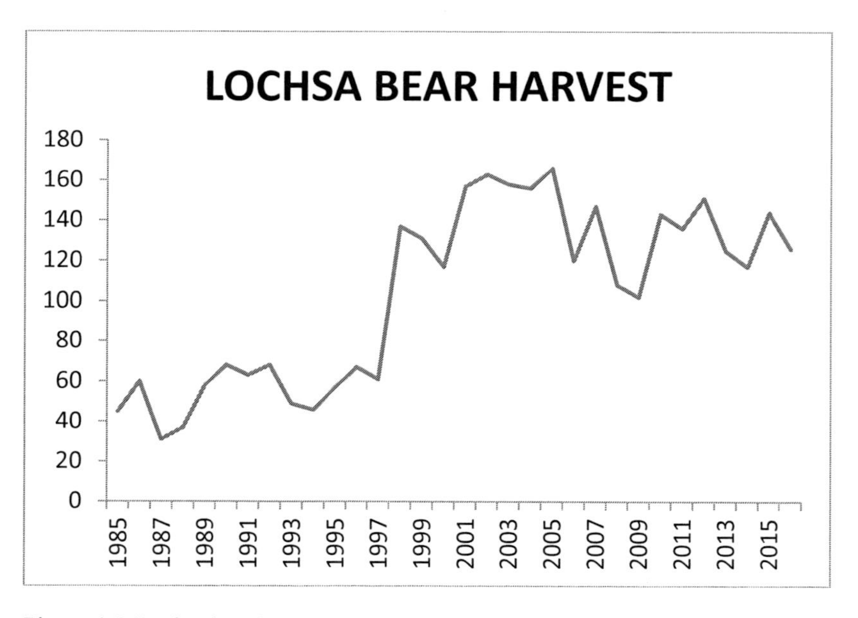

Figure 6-5. Lochsa bear harvests 1985-2016. (Hayden 2018).

The harvest did not reduce mean age of boars being taken from the 1980s to the 2000s (Figure 6-6). There was also no trend in the sow age structure. Idaho Fish & Game considers heavy harvests of boars when the percentage of 5 + year olds in the take is 25% or less. The proportion of the 5+ year-olds in the take from 2004-2012 averaged 33%, suggesting that increased harvest is possible if that 25% management objective is to be reached. The lack of change in age structure of boars over the longer time frame suggests that harvest levels after the institution of the two- bear limit were not affecting the male population. Since black bears are polygynous, there are enough boars present to do the breeding and population trend has not been affected if these data reflect it. Given the lack of vehicle access to a significant portion of the DAU, it is not surprising that overall trend is stable, at least for the period in question. If boars are the major predators on young elk calves, the information suggests that there are plenty of boars around who are aware of the calving concentrations such as Coolwater Ridge to still have a serious effect on calf survival.

Black bear populations are likely regulated through changes in their natural food supply (Garshelis 1994). While there has been a decline in berry-producing shrubs, and habitat conditions for black bears have deteriorated in the Lochsa, the changes have not affected populations as identified through harvest figures. Garshelis (1994) reported densities of black bears from 28 studies across North America ranged from 9-162/100km^2, with the mean density being 33.5/100km^2, (excluding the extraordinarily high estimate for Long Island, Washington). The Coolwater Ridge,Idaho, studies estimated a density of 43/km^2 with only five of the 20 estimates being higher. Not only have the berry-producing shrubs declined on the winter ranges, but as the forest matures the understories will be more shaded and won't have the sun-loving grasses and forbs that are most palatable to bears. Substantial foraging for cambium from conifers further suggested habitat declines (J. J. Beecham, personal communication to Peek, March 29, 2000). However, cuts on the Powell Ranger District and the Western Pacific Timberlands forests in the upper portions of the drainage have produced berry-producing shrubs and herbs and grasses that are bear forages. Historically, wildfires that created habitat diversity from unburned forest to open timber to shrublands to recently burned forb/grass stands produced excellent habitat for both blacks and grizzlies.

At Priest Lake north of the Lochsa, Beecham and Rohlman (1994) found selection cuts with abundant forage and some overstory left were preferred habitat. Productivity of huckleberries at higher elevations increases when overstories are opened up. Pinegrass, not known to be especially palatable to elk, was observed to be fed on in spring by black bears. The cuts in the upper Lochsa drainage, while classified as clearcuts, have produced plants that are bear forage. Undoubtedly bears will forage on the edges of these cuts near timber. Selection cuts that leave 220 trees per acre were highly preferred at Priest Lake. These cuts contained dense understories and trees that black bears could climb and this likely applies in the Lochsa as well.

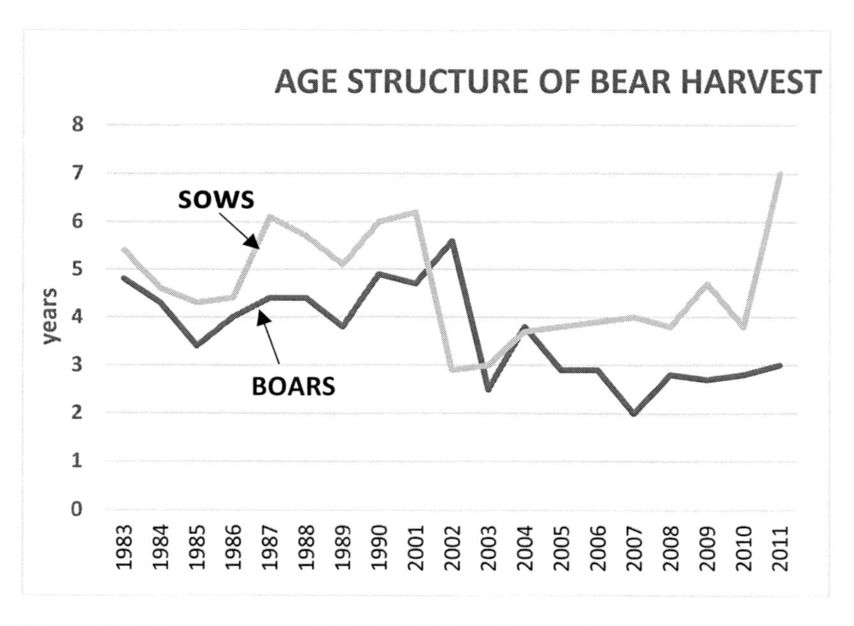

Figure 6-6. Age structure of the black bear harvest in the Lochsa drainage (Hayden2013).

Cougar

The management plans for cougar are predicated on the primary objective of reducing populations which in turn would ostensibly reduce predation on elk. Hunting seasons for cougar in recent years extend from the end of August and go through the winter months until the end of June (Hayden 2014). As with black bears in the Lochsa, a two-lion bag limit and $25.00 nonresident tags has been in force in order to reduce numbers. A minimum harvest objective of 20 lions for the data analysis unit was established in the 1990-1992 management plan. Figure 6-7 shows the cougar take back to 1985. The total take is highly variable but does show a trend with highs in 1997-2000 and then a decline to levels similar to the 1986-1996 period. The take of adult females since 2010 may suggest populations were low with no trend. As with the information on black bears, there is little evidence that the cougar harvest was high enough to either affect the elk population or its trend in numbers. However, this does not mean that local reductions in cougars may have occurred since the Lolo DAU comprises more area than just the Lochsa. Hayden (2014) reported that the harvests since 2010 for the DAU suggest that efforts to reduce the population have been "Moderately effective and that lion numbers have potentially adjusted to decreased prey availability." Also, since 70% of the cougar hunting is done with dogs, hunters may be reticent to turn dogs out on a track in areas where wolves may be present. Numerous incidents of wolves killing hunting dogs have been recorded. Wolves are also known to usurp kills that have been made by cougars (Ruth 2004, Hornocker and Negri 2010).

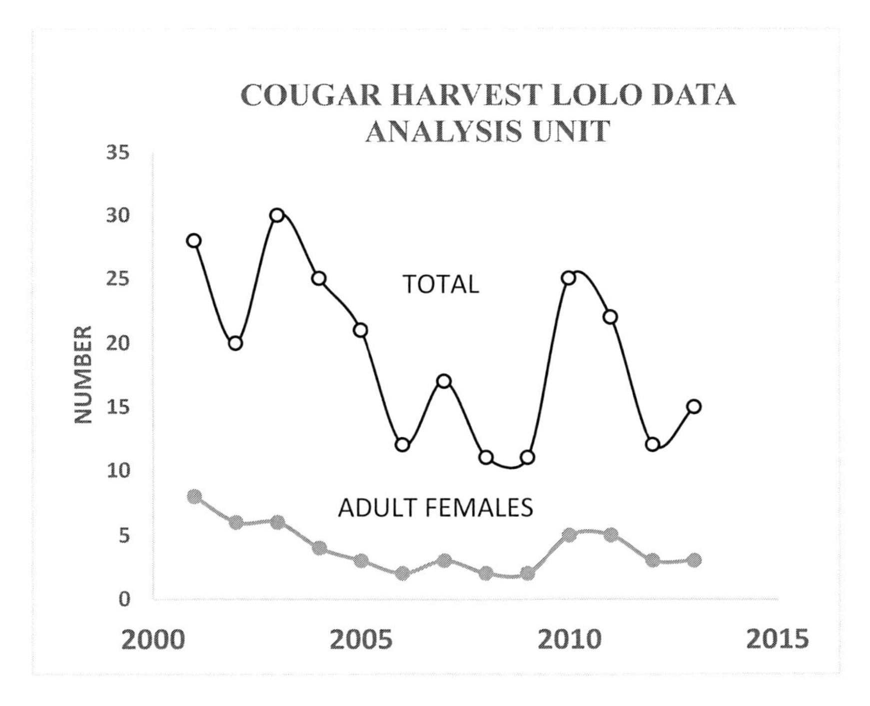

Figure 6-7. Cougar harvest in the Lolo data analysis unit that includes the Lochsa and the North Fork Clearwater (Hayden 2014).

Wolves

The grey wolf, introduced into the Frank Church River-of-No-Return Wilderness in 1995-1996 had expanded its populations into the Lochsa-Selway River drainages by 1997. The introduction of 41 wolves into Yellowstone National Park resulted in an increase to over 170 and then a stabilization starting in 2009 at around 100 or slightly less through 2015 (Smith et al 2016). These conditions were approximated in the Idaho recovery area with 35 wolves being introduced. While the records for Idaho are not as detailed, it is important to note that this restoration process released wolves during a severe winter which would favor the predator over the prey. Also, the prey base was at high density, consisted of a high proportion of older animals, and was not very productive. The severe winter conditions and a highly vulnerable elk population in both release areas benefitted wolves. As a result, wolf populations increased rapidly and elk populations declined. From the standpoint of wolf conservation, these conditions could not have been better. Now, the Yellowstone record shows that the wolf population inside the park has stabilized (Boyce 2018) and the elk population is increasing (memorandum 25 January 2018, Northern Yellowstone Cooperative Working Group. National Park Service, Yellowstone National Park, Mammoth, Wyoming).

Approximately 35 wolves in seven wolf packs had home ranges in 2015 that included the Lochsa River drainage, including some that occupied part of the drainage (Ausband 2016, Hayden 2017). Idaho Fish & Game relies on hunter surveys to locate wolves, since a more thorough estimate would be expensive and time-consuming. Evaluations of this method (Ausband et al. 2014, Rich et al. 2013) indicated that the criteria used did produce a reasonable estimate when 2 to 25 wolves were detected by over three hunters. Documented mortalities of radio-collared elk show that wolves are preying mostly on adult cows in the Lochsa (Table 6-1). Use of trail cameras is now being tried as a means of estimating wolf numbers.

Ausband (2016) reported 19 wolves were killed by control operations and, 23 were taken by harvest in 2015 in Idaho. Approximately two thirds of the sportsman take in Idaho were taken with rifles, with trapping seasons yielding approximately a third of the take over the 2009-2014 period. In winter 2017-2018, 6 wolves were removed from the Lochsa drainage. An ongoing effort to reduce wolf numbers in the upper Clearwater drainage is occurring, without much effect on elk population trends. Between 25 and 30 wolves have been removed from the Lolo Zone during the past 5-6 years, some of these in the Lochsa drainage.

The sum of these investigations show that mortality of adult bulls is primarily from hunting. Adult cow and calf mortality is primarily attributable to predation. Severe winters will abruptly increase mortality of all sexes and ages, if the work in Yellowstone and Scotland applies.

Chapter Seven

Elk Hunting and Numbers

George Hatley, one of the founders of the Appaloosa Horse Club, hunted the Lochsa during the 1950s when elk were abundant. His group of hunters would ride east along the river from the road end which was six miles west of the Lochsa Ranger Station, crossing the river both at Boulder Creek and at Dutch Creek. The trip was about 14 miles from the trail head and perhaps a mile up on Dutch Ridge.

They had a camp with corrals and tent poles on Dutch Ridge that they resurrected each year. George claimed that he never learned to hunt elk because they ran through the camp and were generally around close by so that hardly any stalking was needed. It wasn't unusual to take elk the first day of hunting. Figures 7-1, 7-2, and 7-3 show the country they were in, with the vegetation consisting primarily of shrubs without much conifer. Fall colors of red, pink, purple, yellow, and green spread across that fall landscape could be seen across the Lochsa to the north as well as on the slopes on the south side of the river, spectacular scenery.

One outfitter mentioned to George that he had 55 clients that each got an elk in 1947. Bud Moore (1996) reported that 1,965 hunters checked at a Lolo Pass check station killed 705 elk in 1948. Outfitters spot-packed hunters into camps and returned later to pick them up. Many hunters in those days were prepared to hunt elk on foot, which meant that quartering a carcass, boning the meat, and carrying it out on packboards was often the way elk were hunted and brought out to the trailheads. Those fortunate enough to have mules and horses could quarter their kills and load two quarters on each animal. There weren't many complete carcasses of elk brought out in the days before the roads were built. Even after roads were punched into the side drainages, killing and recovering a carcass meant quartering and often boning out the meat to get it to an access point.

Figure 7-1. George Hatley party crossing the Lochsa River at Dutch Creek In the 1950s. Photo by George Hatley. George Hatley was interviewed by Peek on 10 March 1999.

Figure 7-2. Looking north from Dutch Creek across the Lochsa River. Photograph provided by George Hatley, taken in the 1950s.

Figure 7-3. Trail in Dutch Creek showing the predominance of shrubs. Taken in the 1950s by George Hatley. *See centrefold for this image in colour.*

Outfitters commonly carried hunters into parts of the Lochsa back country that wasn't as frequented by most resident hunters because it was so remote. Moore (1996) explained the advantages of outfitters. The successful outfitter was adept at packing, had knowledge of the country and where best to hunt, and had the trucks and other equipment needed to do the hunting, and tended to be a good conversationalist about the country, the game, their stock, and the hunt. Usually catering to nonresident hunters, outfitters were and are an important component of hunting in the more remote back country areas in the Clearwater as well as the other areas in Idaho

Herman Kuykendall outfitted hunters out of Boulder Creek from 1964-1978 (personal communication to Peek, 28 March 2018). He divided his hunters into three periods, mid-September, October, and later in November. He had 6 hunters in each group and one group was over in the Selway River adjacent to Boulder Creek. The early hunts were "good" but the mid-October hunt, when the full moon was out, were not as good. His hunters each had a horse, guides had three horses plus nine pack mules to constitute the remuda of 18 head of stock in each camp. Many of his hunters were satisfied to take elk of either sex. Hunters who were also after bears were quite successful, with 20 hunters taking 19 bears one year. Kuykendall considered the adult boar black bear to be the predator on elk calves and remembered seeing sows feeding amongst a group of elk.

Hunting Seasons

The first big game seasons consisted of closing hunting from February 1 to July 1 in 1864 (M. Morache. 1974. unpublished report on history of big game, IDFG, Boise). In 1887, the season was further shortened to September 1 through November 30. In 1899, a bag limit of 2 elk was reported, which was reduced to 1 elk in 1912. Parsell (unpublished manuscript, US Forest Service), reported that elk were rarely observed in the Lochsa in 1900, being confined to the eastern end of the drainage, and remained scarce through 1917. However, there are conflicting reports. William E. Parry of Kooskia, resident since 1890 and instrumental in creating the Selway Game Preserve in 1919 (Figure 7-4), reported that approximately 100 elk wintered in the "Deadman Elk Concentration Area" around 1898, where he ran a trapline (Fry and Parsell 1938). In 1911, the elk season was closed in Bonner, Kootenai, Shoshone, Latah, Nez Perce, Clearwater and Idaho counties, which includes the Clearwater River drainage and the Lochsa. State Game Warden Gowen reported that the closures appeared to be effective in increasing elk, and recommended that areas remaining open, primarily near Yellowstone National Park, be changed to bull only from September 15 to November 30.

Space (1964) reported that a hunter killed an elk in Fourth of July Creek on the North Fork of the Clearwater in 1919, when the season in the Clearwater drainage was opened from October 1-November 15. The 1922 Idaho Fish and Game Biennial report included a statement from a Mr. Jones: "During a five-week pack trip in the Lochsa and Selway River, we counted as high as 12 elk and 14 deer in one day's travel on foot from one camp." In the summer of 1924, Space (1964) saw "numerous elk on Weir Creek and around Jerry Johnson and Colgate Hot Springs. Elk were plentiful in Stanley Creek and Lake Creek and the head of Boulder Creek." These are all drainages that enter the Lochsa.

The Carlin Party of 1893 started out from Kendrick, Idaho, on the 18th of September, arriving at Bald Mountain on the North Fork of the Clearwater and the Lochsa Divide on or about September 25 (Hamilton 1997). The purpose of their trip was to hunt, but fall storms, first rain, then snow, hindered their efforts. The did take a nice bull elk and saw plenty of elk sign, especially around the licks (Colgate Licks) near Jerry Johnson's cabin and hot springs. It is evident that elk were plentiful enough by the late 1800s to warrant hunting them. Lieberg (1899) mapped regenerating timber and recent burns in the vicinity of these licks, as well as mature cedar forest along the river (Figure 3-2).

These kinds of casual observations from people traveling and working in the area constituted the major source of information on the elk population in the 1910-1930 period. They did provide some evidence that elk were increasing but the causes and extent of the increase were hardly considered.

***

SELWAY GAME PRESERVE. SECTION 2848 ON PAGE 64 OF FISH AND GAME LAWS OF 1921-22. SECTION 2848.

BOUNDARIES. The area embraced within the following described boundaries is set aside and designated as a game preserve to be known as the Selway game preserve:

> Beginning at the confluence of the Old Man creek with the Locksaw river; thence up said Old Man creek and the west fork thereof to the summit of the Craggs mountains at Snow hill; thence on and along the crest of said Craggs mountains to a point opposite the headwaters of Rhoda creek; thence down said Rhoda creek to its confluence with the north fork of Moose creek; thence down said north fork of Moose creek to its confluence with the Main Moose creek; thence on and up said Moose creek to the line between the state of Idaho and the state of Montana in the Lost Horse pass; thence northerly on and along said line between the said states of Idaho and Montana to the Deep creek pass; thence in a direct line to Little Sand lake at the headwaters of Storm creek; thence down Storm creek to its confluence with White Sand creek; thence down White Sand creek to its confluence with the Locksaw river; thence down the south bank of the Locksaw river to the mouth of Old Man crek and point of beginning, the area embraced therein being part of Idaho county or any county created therefrom within the state of Idaho.
>
> Sec. 2849. All Game Animals Protected. It shall be unlawful for any person or persons at any time to hunt, trap,capture, pursue or kill game animals of any kind or description whatever within the limits or boundaries of the Selway game preserve.
>
> Sec. 2850. Fur-Bearing Animals. It shall be unlawful for any peson or persons at any time to hunt, trap, capture, pursue or kill within the limits or boundaries of the Selway game preserve any of the following fur-bearing animals, to-wit: Beaver, otter, marten, fisher, fox or mink.
>
> Sec. 2851. Predatory Animals. Mountain lion, bear, lunx, timber or prairie wolf, wolverine, coyote and wild cat may be destroyed or exterminated within said Selway game preserve by game wardens or persons authorized by game wardens to destroy such predatory animals.

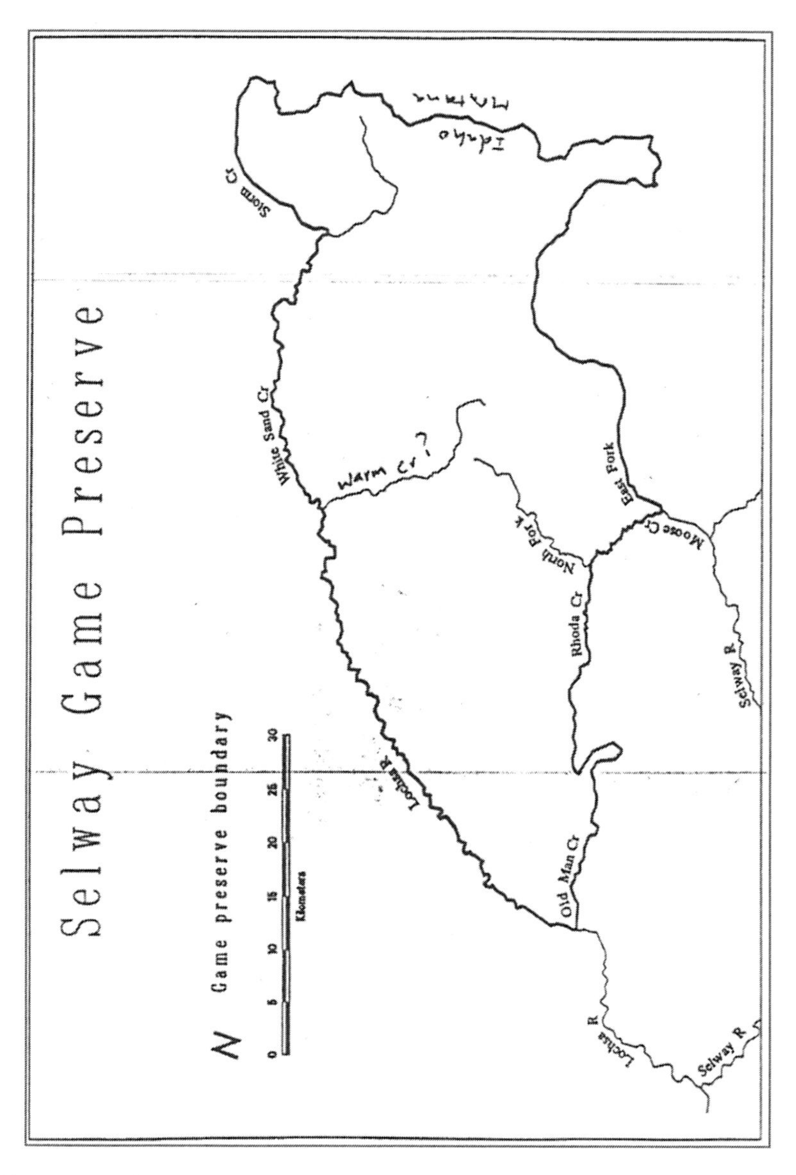

Figure 7-4. Selway Game Preserve 1919-1951.The outline shows the Lochsa River, White Sand Creek, and Storm Creek on the north, the Idaho-Montana border on the east, the East Fork of Moose Creek and Rhoda Creek in the Selway River drainage on the south, and Old Man Creek in the Lochsa drainage. Original spelling included.

However, hunting was eliminated in the Selway Game Reserve and predator control was ongoing, directed at coyotes, cougars, black bears, and bobcats because the wolf and grizzly had been virtually extirpated earlier.

In the early days, management activities centered around control of predators and poachers. C.L. Monroe, assistant chief deputy game warden wrote of an illegal tusk hunting incident on the Idaho-Montana line in the Lochsa in 1914 (Monroe 1914). Hunters reported poaching and investigations found four elk carcasses on the Idaho side and four on the Montana side, with only the teeth (the upper canine or tusk) taken. The investigating party consisted of the game warden, an Idaho county sheriff deputy, a professional packer, and one Forest Service ranger. The group met a number of hunting parties, both whites and Indians, some of which were going to spend the winter hunting cougar (bounty, $25.00).

The territorial legislature in 1887 established an either sex hunting season on elk in the state to run from September 1 through December 31, 122 days (Table 7-1). After statehood, the legislature met biennially, and hunting seasons were established by law for the two-year period. In 1908, a September 15-December 31 season of 91 days was enacted. In 1917, a September 16-November 30 season was established for the state, but in Idaho County, the season was November 1-30, and calves were protected. From 1921 to 1933, a season from October 1 through November 15 of 46 days was in force, and calves were protected. Through the rest of the 1930s, the season was from October 5 to November 10, or 36 days. The legislature, through the 1887-1939 period essentially set a pattern of hunting elk in October in Clearwater and Idaho Counties, which included the Lochsa, for either sex, with calves protected. On November 8, 1938, a citizens initiative was approved to establish a state fish and game commission, which was approved by the legislature as the Idaho State Fish and Game Commission Act, Initiative Measure 1939. The commission was given authority to establish regulations after that. The department was established in 1938 as well.

Through the 1940s and early 1950s, the hunting season philosophy was designed to accommodate the tradition of hunting either sex, which also fit the objective of attempting to keep elk numbers at levels below which the forage supplies would be affected by heavy browsing. This period reflected excellent numbers, not too many hunters, and access primarily by trail into the hunting areas. The management philosophy was typical for management of big game across the west at that time. Mule deer numbers, as well as elk numbers in some areas, were high and heavy browsing and winter die-offs were common occurrences. Game managers were trying to reduce numbers in order to curb the die-offs and ensure that browse supplies were not deteriorated.

Table 7-1. Summary of elk hunting seasons in the Lochsa River drainage.

YEAR	OPEN	CLOSE	SEASON LENGTH	SEX	REMARKS
2016	Oct 9	Nov 2	24	antlered	Lolo Zone B-Tag capped at 1600
2015	Oct 9	Nov 2	24	antlered	Lolo Zone B-Tag capped at 1600
2014	Oct 9	Nov 2	24	antlered	Lolo Zone B-Tag capped at 1600
2013	Oct 9	Nov 2	24	antlered	Lolo Zone B-Tag capped at 1600
2012	Oct 9	Nov 2	24	antlered	Lolo Zone B-Tag capped at 1600
2011	Oct 9	Nov 2	24	antlered	Lolo Zone B-Tag capped at 1600
2010	Oct 9	Nov 2	24	antlered	Lolo Zone B-Tag capped at 1600
2009	Oct 9	Nov 2	24	antlered	Lolo Zone B-Tag capped at 1600
2008	Oct 9	Nov 2	24	antlered	Lolo Zone B-Tag capped at 1088
2007	Oct 9	Nov 2	24	antlered	Lolo Zone B-Tag capped at 1088
2006	Oct 9	Nov 2	24	antlered	Lolo Zone B-Tag capped at 1088
2005	Oct 9	Nov 2	24	antlered	Lolo Zone B-Tag capped at 1088
2004	Oct 8	Nov 2	25	antlered	Lolo Zone B-Tag capped at 1088
2003	Oct 9	Nov 2	24	antlered	Lolo Zone B-Tag capped at 1088

2002	Oct 9	Nov 2	24	antlered	Lolo Zone B-Tag capped at 1088
2001	Oct 9	Nov 2	24	antlered	Lolo Zone B-Tag capped at 1088
2000	Oct 9	Nov 2	24	antlered	Lolo Zone B-Tag capped at 1088
1999	Oct 9	Nov 2	24	antlered	Lolo Zone B-Tag capped at 1088
1998	Oct 9	Nov 2	24	antlered	Lolo Zone B-Tag capped at 1088
1997	Oct 10	Nov 3	24	antlered	mountain tag
1996	Oct 10	Nov 3	24	antlered	mountain tag
1995	Oct 10	Nov 3	24	antlered	mountain tag
1994	Oct 10	Nov 3	24	antlered	mountain tag
1993	Oct 10	Nov 3	24	antlered	mountain tag
1992	Oct 10	Nov 3	24	antlered	mountain tag
1991	Oct 10	Nov 3	24	antlered	
1990	Oct 3	Oct 28	26	antlered	
1989	Oct 4	Oct 29	25	antlered	
1988	Oct 5	Oct 30	25	antlered	Sep 15-Oct 24, east of USFS roads
1987	Oct 7	Nov 1	26	antlered	568 &500, south of US 12.
1986	Oct 1	Oct 26	26	antlered	Sep 15-Oct 26, east of USFS roads
1985	Oct 2	Oct 27	26	antlered	568 &500 south of US 12
1984	Oct 2	Oct 27	26	antlered	Sep 15-Oct 28

1983	Oct 5	Oct 30	26	antlered	
1982	Oct 6	Oct 31	26	antlered	Sep 15-Oct 31
1981	Sep 30	Oct 25	26	antlered	Sep 16-Oct 26
1980	Oct 1	Oct 26	26	antlered	Sep 17-Oct 26
1979	Oct 3	Oct 28	26	antlered	Sep 12-Oct 28
1978	Sep 27	Oct 22	26	antlered	Sep 13-Oct 22
1977	Sep 28	Oct 23	26	antlered	Sep 14-Oct 23
1976	Sep 29	Oct 24	26	antlered	Sep 15-Oct 28
1975	Sep 15	Oct 26	31	antlered	Sep 15-Oct 28
1974	Oct 5	Nov 3	29	either sex	Sep 14-Oct 13
1973	Oct 6	Nov 4	29	either sex	Sep 15-Oct 14
1972	Oct 7	Nov 5	29	either sex	Sep 16-Oct 15,
1971	Oct 2	Oct 31	30	either sex	Sep 18-Oct 17
1970	Oct 3	Nov 1	30	either sex	to Lolo Pass
1969	Oct 4	Nov 2	30	either sex	
1968	Oct 5	Nov 3	30	either sex	
1967	Oct 7	Nov 5	30	either sex	
1966	Oct 1	Oct 30	30	either sex	

1965	Oct 2	Oct 31	30	either sex	
1964	Oct 3	Nov 1	30	either sex	
1963	Sep 28	Oct 27	30	either sex	
1962	Sep 29	Nov 15	48	either sex	
1961	Sep 29	Nov 15	46	either sex	
1960	Oct 1	Nov 15	46	either sex	
1959	Oct 3	Nov 15	44	either sex	
1958	Oct 4	Nov 2	30	either sex	
1957	Oct 1	Oct 31	31	either sex	
1956	Oct 1	Oct 31	31	either sex	
1955	Oct 1	Oct 31	31	either sex	
1954	Oct 1	Oct 31	31	either sex	
1953	Oct 1	Oct 31	31	either sex	
1952	Oct 1	Oct 31	31	either sex	
1951	Oct 1	Oct 31	31	either sex	
1950	Oct 1	Oct 31	31	either sex	Sep 20-Oct 31, Selway Game Preserve
1937	Oct 5	Nov 10	36	either sex	1000 permits
1935	5 Oct	Nov 10	36	either sex	

1933	Oct 5	Nov 10	36	either sex	calves protected
1929	Oct 1	Nov 15	46	either sex	calves protected
1927	Oct 1	Nov 15	46	either sex	calves protected
1925	Oct 1	Nov 15	46	either sex	calves protected
1923	Oct 1	Nov 15	46	either sex	calves protected
1921	Oct 1	Nov 15	46	either sex	calves protected
1917	Sep 16	Nov 30	75	either sex	calves protected, Nov1-30, in
1908	Sep 15	Dec 31	91	either sex	Idaho County
1887	Sep 1	Dec 31	122	either sex	hide, horn hunting prohibited
					taking more than 1 ibex illegal

______

Important changes in management philosophy came with the change in game managers in the Lewiston office of Fish and Game which manages the Clearwater drainage (G.R.[Sam] McNeill, personal communication to Peek, 24 May 1995). Elmer Norberg felt that elk hunting seasons should be short, so harvests would be light. The objective was to keep as many elk in the population as possible to keep browsing pressure on the shrub forage high. This recognized that plant succession was proceeding and an effort to delay it was the objective. However, when McNeill succeeded Norberg, hunting seasons were revised to be longer so more elk would be taken, presumably so a lower population level would trigger a compensatory response in reproduction and survival, leading to a more thrifty population. This strategy continued until cow hunting was eliminated in 1975, when Governor Andrus (1931-2017) urged the Fish & Game Commission to quit hunting cows, undoubtedly at the urging of hunters and outfitters (personal communication from Governor Cecil Andrus to Peek, 16 April 1998). Shortly afterward, emphasis on reducing predators became an important objective after Schlegel's work in the late 1970s identified predation as the major factor in reducing calf survival.

Regardless, it had become apparent by the early 1970s to hunters and managers alike, that the low elk numbers required a change in management. Also, at existing population levels the highly palatable browse plants such as redstem ceanothus still showed heavy use, and the taller growing shrubs were out-of-reach of elk. In 1975, unlimited hunting of either sex elk across the state and in the Lochsa was curtailed in favor of hunting only antlered bulls, except in the heavily forested Idaho Panhandle units further north. In 1971 preceding the general change, mid-September seasons for antlered bulls in the wilderness and least accessible areas were established to allow back-country hunters to hunt elk in the rut. The opening dates for the entire Lochsa were moved into late September in 1975, but this lasted through 1978, when the season went to the first part of October. Hunting season lengths had changed from 30 days in the 1960s to 26 days in 1976 and thereafter. In 1988 the September 15th opening on the south side of the river was abandoned and the entire unit was opened the first weekend in October. Low bull survival resulted in the next change, which entailed starting the season on the 10th of October, after the rut ended. In 1992, in a further effort to address the increasingly apparent low bull survival problems that were occurring in hunted elk herds throughout the west including Idaho, back-country units were designated to be hunted with a "mountain tag". The Idaho Panhandle, including the St. Joe River drainage and further north, was separated into a special "Panhandle tag", requiring hunters to choose that area or the rest of the state. Idaho's elk regulations thus were attempts to restrict hunter opportunity in an effort to increase bull survival and to recognize differences in accessibility and forest conditions in different areas of the state.

Hunter numbers in Idaho increased from about 69,000 in 1982 to over 105,000 in 1995 and then declined to average 86,618 over the 2001-2014 period. Increases in permitted or limited-entry hunts have offset declines in the back-country hunts. In the Lochsa, over the same period, hunter numbers declined (Figure 7-5). Elk harvests in the Lochsa also declined (Figure 7-6). Those who persist in hunting the Lochsa likely include a substantial number who are experienced in locating elk in the drainage.

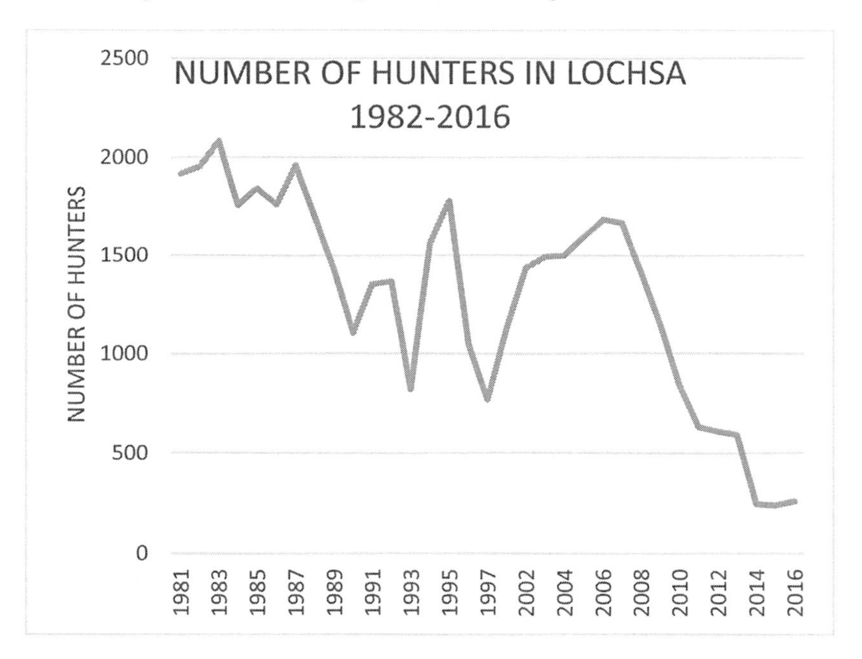

Figure 7-5. Changes in the number of hunters in the Lochsa River drainage, 1982-2016. Idaho Department of Fish and Game.

Formal records of elk harvest for the Lochsa began when the Kooskia check station was initiated in 1947 and other sources became available in 1948. Forest Service had a check station manned by both their employees and Idaho Fish & Game wardens. The data represent a mix of sources, i.e. check stations, mailed questionnaires, hunter report cards, and telephone surveys. Earlier records show inconsistencies but do show patterns. Problems include that many times elk killed either in the Lochsa or Selway

River drainages were combined in the records. Hunter responses to mailed questionnaires provided estimates of elk harvest starting in 1953. In 1959, hunter report cards were issued. Questionnaire and report card formats were changed during the early 1960s. For some years data from check stations at Lolo Pass and Kooskia were combined, so that again harvested elk could not be assigned specifically to the Lochsa unit. Report cards were discontinued in 1979 and a telephone survey initiated. Telephone survey data were broken down only to management area so harvest data commonly were not specific to drainages within Unit 12. The telephone surveys provided a means of rapidly canvassing representative samples of hunters with techniques commonly used by contemporary pollsters for a variety of purposes. Hunter check stations at Lolo Pass and Kooskia were discontinued in 1985. Mandatory hunter reports were instituted with success in 1998.

High elk harvests in the Lochsa occurred from 1948-1961, peaking in 1959. Harvests then fluctuated at lower levels and then declined abruptly in 1997 after the dieoff of the preceding winter (Figure 7-6), coinciding with the observed trend in populations. These data are the checked or reported harvest only and so do not represent the total harvest. Indian harvests and other human-caused mortalities (poaching, road kill) are not included in the data. Our assumption is that the surveyed harvest consistently reflects trends in the total harvest and the distribution of that harvest. Distribution of elk harvest over the 20-year period was similar when compared to the highest kill in the 1971-75 period.

If data obtained by these various surveys reasonably reflect the numbers and distribution of elk harvest, then elk were killed mostly in areas with few roads and little timber harvest. For the period pre-1971 through 1990 (Figure 7-7), the highest elk harvest occurred north of the river in areas with few roads ($<0.6 km/km^2$). Areas with the second highest elk harvest were south of the river (again with few roads) prior to the change in hunting regulations in 1975 (switch to bulls only). After 1975, areas north of the river with high road densities accounted for more of elk harvested than south of the river. Low bull:cow ratios in the 1980s, although declining on both sides of the river, were much lower on the south side, which includes the Selway-Bitterroot Wilderness, and apparently reflect the greater harvest of mature bulls in these sparsely roaded, non-logged areas. The reduced season south of the river recommended by Idaho Fish and Game was finally instituted five years later. Cow:calf ratios also were generally lower south of the Lochsa River beginning in the late 1970s (Figure 7-8).

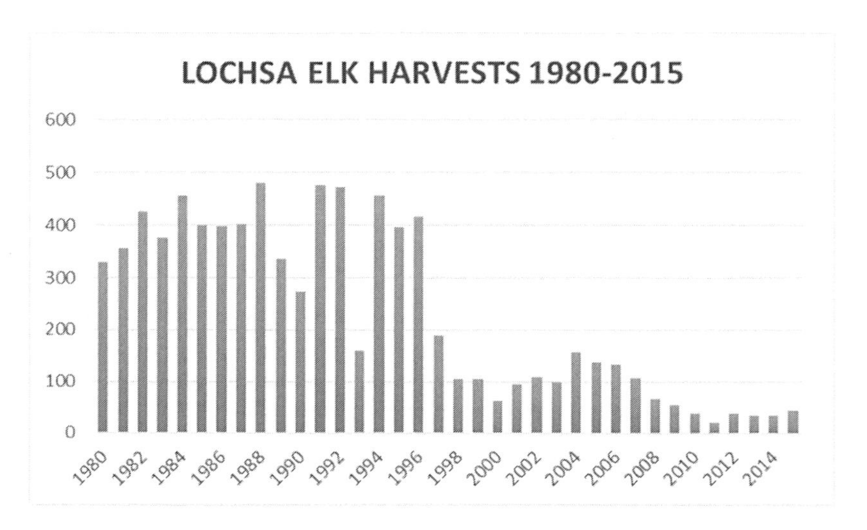

Figure 7-6. Elk harvests in the Lochsa River drainage, 1980-2015. Data available from Idaho Department of Fish & Game elk reports.

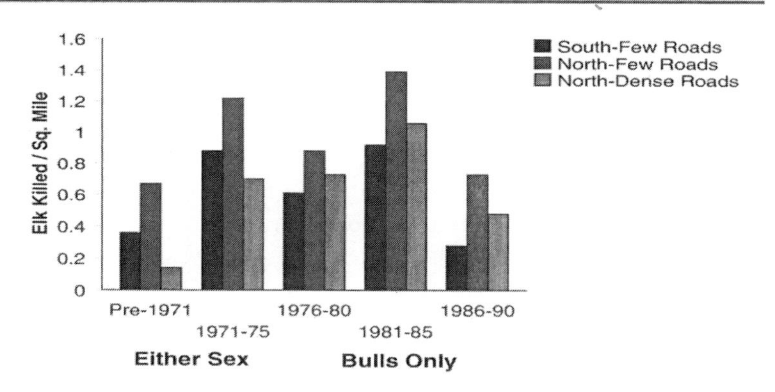

Figure 7-7. Distribution of elk harvest in the Lochsa drainage related to roads through 1990.

Of the variables summarized (road density, proportion of area logged, proportion of area burned by wildfires, and proportion of area burned by planned ignitions), only planned ignitions had a significant relationship with elk harvest but this relationship accounted for only 12% of the variation

(Figure 7-8). This relationship was most evident beginning five years after the initiation of the prescribed burning program (when at least half the total acreage had been burned) and continued significant until 1985. By that time very little area was being burned by the USFS.).

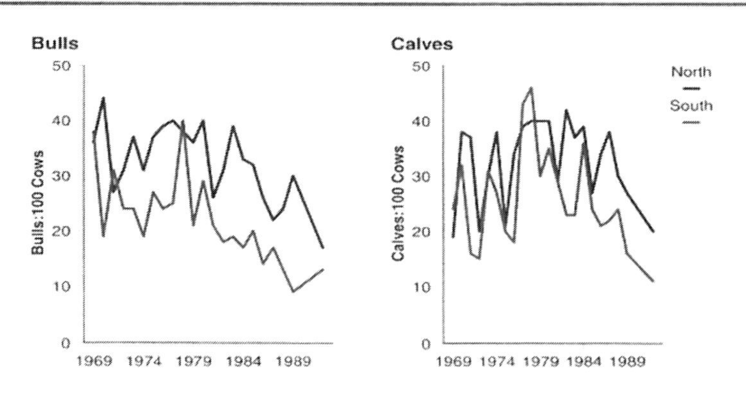

Figure 7-8. Comparison of age and sex ratios in the Lochsa drainage north and south of the river.

Fish and Game followed up on these observations for the five-year period, 1986-91. Unsworth et al. (1993) found that hunting during the September-October period accounted for 86% of elk mortality in the Lochsa. They calculated annual survival rates for cows as 89% ($\pm$9%) and for bulls (both yearlings and older) as 60% ($\pm$6%). Post-season bull:cow ratios in the western portion of the drainage were 14:100 and in the eastern portion 35:100. The western portion was extensively roaded and logged in the upper parts. Modeling the relationship between elk mortality, and such management and habitat variables as open or closed road densities, topographic roughness, plant community (closed timber, open timber, shrub field), and hunter densities, Unsworth et al. (1993) found that the probability of mortality significantly increased as open road densities increased and as hunter densities increased but declined as topographic roughness increased. They concluded that patches of timber in extensively roaded areas may increase bull elk vulnerability. This is because hunters can readily access these stands and, because these stands also are generally isolated and small, hunters can more efficiently hunt them. Elk select for timbered habitats in early fall to seek succulent forage, escape the heat, and for security.

In roaded areas (> 2.5 km/km^2) of the Lochsa, elk preferred timbered habitat but in these areas they were twice as likely to be killed by hunters than in unroaded areas (< 0.3 km/km^2, Unsworth and Kuck 1991). Bull elk in unroaded areas preferred open timber habitat. Survival rates for bulls in roaded areas were 44% for yearlings and 41% for adults. In unroaded areas, yearling survival rates were 79% and 78% for adult bulls. Although the area of timbered habitat in roaded and unroaded areas was similar, timber in roaded areas was more fragmented.

The harvest record from 2000-2016 shows an average of 78 (range from 22-158) bulls taken during the hunting season. This tracks the decline in elk in the drainage but reflects declining hunter interest, conditions during the hunt, locations of bulls more in coniferous timber where they would be less easily observed, and removal of general season hunting with firearms during the rut.

Elk Numbers

Elk counts in the Lochsa began in 1936 and continued intermittently until 1969 when regular annual winter censuses became standard practice. Although early counts must be viewed as minimum estimates at best, and data until 1969 were sporadic, there's general agreement that elk numbers increased following the 1934 fires. Elk apparently continued to increase into the 1950s, but a steep decline began in the early 1960s with extensive browsing evident and opening up of the drainage to timber harvest and completion of the highway. During the 1970s when bulls-only hunting was instituted, elk numbers increased but another population decline is evident since. Ability to estimate elk population numbers improved significantly in 1985 when a “sightability” model was incorporated into census estimates for the Lochsa elk herd (Samuel et al. 1987). The model provides an estimate of the number of elk not seen during the census (based on sex, group size, vegetation cover, and snow cover) and provides confidence intervals on population estimates. Areas censused for "sightability" estimates are chosen randomly within the drainage whereas the earlier censuses tried to cover the entire winter range using the experience of those involved as to what that area would consist of. Census estimates obtained beginning in 1985 indicate that actual numbers of elk may be > 40% higher than the number of elk observed. Mature bulls particularly were under-represented among elk observed, not unexpected since they may winter in deeper snow areas, in heavier cover, and in smaller groups or lone individuals. Table 7-2 summarizes corrected census and proportions of age and sex for the 1980-2010 period.

The downward trend in elk population estimates from the 1980s through the 2000s can be estimated from the surveys. Mean corrected population estimate for the 1981-1989 period was 4690 elk (range 3558-6104). Three population estimates for 2002, 2006, and 2010 show a mean of 1470 elk, or 31% of the mean estimate for the 1980s. The proportion of calves in the population for the 2000s was 22% of the mean estimate for the 1980s. Barring severe winter dieoffs, calf losses are the major mortality which in turn eventually reduces recruitment of breeding-age individuals. However, these dieoffs also involve adults, especially older cows and bulls. When that happens the dieoffs will cause an abrupt, steep drop in numbers that can be interpreted as a threshold response. Also, wolf predation on adult cows contributed to the decline and likely accelerated it. The continuing low calf survival caused a less abrupt trend but was still appreciable when considered over decades. The trend downward that was attributable to lower calf survival was caused by predation, but the die-offs were related ultimately to limited forage supplies, even as predation was the immediate or proximate cause. Still, it is not well understood just why black bears are the main predator involved in calf mortality, except that bear populations continue to be high in the Lochsa as well as elsewhere in northern Idaho.

Table 7-2. Census of elk in the Lochsa River drainage, 1980-2017. Idaho Fish and Game.

Year	Total	Cows	Bulls	Yrlng	Raghorn	Branch Antlered	Calves
1980	2974	1700	603	214	148	389	656
1981	3558	2014	529	257	78	273	595
1982	5296	3250	981	406	206	574	1048
1983	4505	2704	924	303	339	621	842
1984	6104	3624	1006	386	268	620	1355
1985	4775	2852	968	314	349	653	857
1986	4913	3060	1034	222	379	812	793
1987	4615	2906	703	240	250	464	908
1988	4550	2956	738	271	286	466	855
1989	3893	2466	802	290	344	511	626
1992	3452	2515	550	132	137	417	381
1994	3315	2415	446	144	148	301	326
1995	3852	2754	465	135	212	329	599
1997	2667	2060	425	264	327	264	327
2002	2048	1281	422	169	102	253	343
2006	1658	978	475	132	180	343	196
2010	705	534	134	9	50	125	38
2017	484	271	159	36	94	122	51

Three winters were severe enough to cause major dieoffs in the Lochsa. The 1948-49 winter was the most severe (Chapter 5), but dieoffs occurred during the 1978-79 winter and the 1996-97 winter. Census information prior to the use of sightability estimates is unknown as to its accuracy, although they will invariably be lower estimates than what occurred at the time. The earlier estimates likely do reflect trends (Figure 7-9). The 1996-97 dieoff is reflected in an approximate loss of 30% of the population if the figures for 1995 and 1997 can be compared. Calves seen in 1997 were approximately 50% of the 1995 observations. Cows were approximately 25% fewer in 1997 than in 1995. The observations of the different age classes of bulls likely reflects different distributions in forested areas and different ability to classify individuals between the years. Additionally, snow accumulation was different for the 1995 and 1997 winters with 1158 cm (456 in) 1995 and 1692 cm (666 in) in 1996-97 for the November-January period. This would be the cause of the dieoff. The 1978-79 winter had 2098 cm (829 in) of snowfall for the November-January period, which would also precipitate a dieoff (Figure 7-10). These estimates of the loss are low because the censuses were done in mid-winter while much of the loss would come in late winter and early spring.

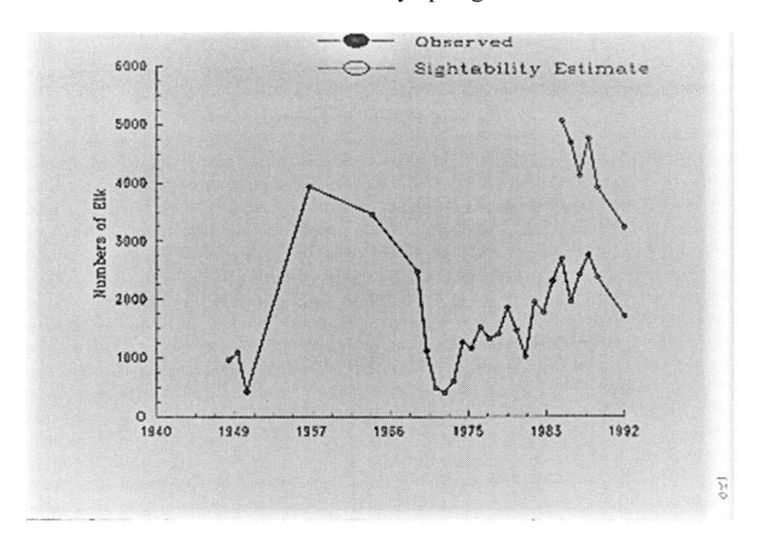

Figure 7-9. Elk population estimates from 1948-1992 using fixed-wing aircraft by Idaho Fish & Game. The estimates using "sightability" from 1984-1992 reflect the inevitable low number observed and provide an estimate of those not seen. The observed estimates likely reflect the general trend but are not an accurate estimate of numbers.

Figure 7-10. Deep snows such as these will limit foraging and movement and precipitate dieoffs of elk in the Lochsa.

CHAPTER EIGHT

HABITAT RELATIONSHIPS

Food Habits

Elk foraging patterns will depend upon what is available, what is most palatable, and what can be most easily digested. Woody plants are a large part of the diet in the Lochsa for most of the year although forbs and grasses will be taken in spring through early fall when they are available. The earliest investigations of foraging patterns were done by Vernon Young and Leslie Robinette (1939) on the Wounded Doe Licks drainage on the Selway. One study site was an open browse area of eastern exposure at 5000 feet (1500 m) elevation. The other plot was established on a western exposure. The studies were done during the summer of 1938. These sites had been burned in 1910 and had been subject to extensive browsing by deer and elk. Table 8-1 shows the summary of 15 weekly examinations of utilization of the vegetation on both sites. Shrubs were more heavily used than flowering forbs and grasses, but ground cover percentages suggested that very few shrubs were available. However, mountain maple, serviceberry, current, willow, elderberry, mountain ash, and huckleberry were the woody plants that were heavily used. The composition of the study sites plus the proximity of the sites to the Wounded Doe Licks suggests that these sites would have been subject to extensive browsing by elk.

Table 8-1. Estimates of utilization at two sites on the Wounded Doe Licks in the Selway River by Young and Robinette (1939). Scientific names are listed in Appendix A.

Species	Site One % Ground cover[1]	% Utilization	Site Two % Ground cover	% Utilization
Hoary tansyaster	-	10	-	10
Eaton's asteri	-	10	-	10
Broadleaf arnica	1	5	-	
Lady fern	-	20	-	
Fireweed	-	10	-	
Spring beauty	1	10	-	
Rocky Mountain larkspur	-	5	-	
Blanketflower	-	15	-	
Fringed hawkweed	-	20	-	
Tall fringed bluebell	-	30	-	25
Varileaf penstemon	-	25	-	
Littleflower penstemon	-	55	-	20
Arrowleaf ragwort	-	15	-	
Star false Solomon's seal	-	10	-	
Meadow rue	-	10	-	
Sitka valerian	-	10	-	
Mountain maple	-	65	-	
Serviceberry	-	40	-	15
Shineyleaf ceanothus	3	25	12	10
Menziesia	5	15	-	
Bitter cherry	-	60	-	
Stream currant	-	25	-	
Sticky currant	1	45	-	45

Thimbleberry	-	10	1	15
Scouler willow	1	60	-	40
Rocky Mountain elder	-	95	-	100
Mountain ash	-	30	-	40
Shineyleaf spiraea	-	10	-	15
Common snowberry	-	35	-	
Thinleaf huckleberry	8	15	1	45
Geyer's sedge	20	5	10	5
Grasses	15	10	15	5

[1]Species with no ground cover were less than trace.

Hash (1973) used two methods to obtain year-long food habits of elk in the Lochsa. Rumen samples were collected from 139 elk mostly during the 1971-72 winter from kills by tribal hunters. Other rumens were collected from fall hunter kills. Captive elk supplied rumen contents through use of a trocar to pump contents from rumens using techniques commonly applied to domestic ruminants (Follis and Spillett 1972). Examination of 143 feeding sites was the other method used to assess food habits, which included intensive examination of a recently used site for evidence of plant use, plus an assessment of plant composition in the vicinity of the site. A comparison of the two methods indicated that they produced similar values. Most of the feeding sites were in open areas or brushfields because of the increased opportunity to locate animals as compared to forested areas. For the entire study period including all seasons and combining the results of both methods, browse constituted 64% of the diet, grasses and grass-likes 25%, and forbs, lichens, and mushrooms being the remainder. Major differences in the observed winter diet between the two winters reflected the relative severity of each winter (Table 8-2). The 1971-72 winter was severe with deep snows and shrubs constituted 92% of the diet while the next winter with lower snow depths shrubs comprised 69% of the diet. Conifers that were a large part of the diet during the severe winter showed a marked decrease in the diet for the milder winter. Snowbrush ceanothus, Scouler willow, and redstem ceanothus were the major items in the diet during the mild winter. Redstem ceanothus, western red cedar, and Douglas fir were the major items in the diet during the severe winter. Creeping barberry, pachistima, and roses became more important in the diet as snows melted and these shorter forages became available. No observations of elk attempting to paw down through the snow pack to reach lower forages were made. Hash (1973) reported seeing many animals in very poor condition during the 1971-72 winter.

Spring diets increasingly changed from browses to grasses and sedges as temperatures and soils warmed and elk were able to use more areas. Grasses increased in the diet from 25% in March to 65% in June. Forbs increased steadily throughout the summer diet but redstem and snowbrush ceanothus remained important as did the grasses. Beargrass was taken in moderate amounts, suggesting that elk were using higher elevations that would include parks and burns in spruce-fir habitat types. Fall diets showed the transition from higher summer forb use to more use of shrubs. Grasses remained steady at about 13% of the observed diet. Pinegrass and sedges were the most commonly used, indicating elk were moving to lower elevations where these species were more abundant.

Nowlin (1974) took advantage of the harvest of elk by the Nez Perce Tribe in winter plus live-trapped elk to obtain rumen contents for trials to determine the digestibility of forages. Rumen contents were collected in March, and the inoculum fluid containing the microorganisms that do the digesting was inserted into vacuum flasks and introduced to twigs collected from shrubs known to be palatable to elk. A dairy cow supplied rumen inoculum as a comparison. Shrub collections came from recently burned sites as well as unburned sites. Digestibility of mountain maple, ocean spray, redstem ceanothus, serviceberry, Douglas fir, and willow was estimated (Nowlin 1974). Mountain maple (36%) was the highest in digestible dry matter with Douglas fir (31%) next. Redstem ceanothus (29%) and serviceberry (30%) were next, with willow (25%) last among the species that were preferred forage. Ocean spray (18%) was last and is not a palatable forage. Burned versus unburned values showed little differences except for willow being higher on burned sites. Dry matter digestibility using the dairy cow inoculum and captive elk fed on alfalfa was consistently lower than for the wild elk, probably because the microorganism component in the captive animals was different.

Douglas fir can be digested when taken at levels where the inhibitory oils that impede digestion can be tolerated. Generally, with the exception of willow, there was little difference in digestibility of palatable forages on the burns when compared to unburned areas. The burns occurred in April 1973, a year prior to the collections. The results imply that forage biomass was more critical than digestibility differences of the important forages, but ocean spray, a species with low preference was also low in digestibility in March. Figure 8-1 shows 6 herbaceous species that are palatable to elk when they develop in snow-free periods.

Table 8-2. Most important forage species for each season after Hash (1973). Names follow U.S. Department of Agriculture Natural Resource Conservation Service. 2018. PLANTS database (http://plants.usda.gov).

	WINTER1971-72[1]	WINTER 1972-73[2]	SPRING 1971 and 1972[3]	SUMMER 1971 and 1972[4]	FALL 1971 and 1972[5]
	JANUARY-FEBRUARY	JANUARY-FEBRUARY	MARCH-JUNE	JULY-SEPTEMBER	OCTOBER-DECEMBER
% BROWSE	91.7	69	43	39	55
	REDSTEM	SHINEYLEAF CEANOTHUS	BERBERIS REPENS	SHINEYLEAF CEANOTHUS	SCOULER WILLOW
	WESTERN RED CEDAR	SCOULER WILLOW	DOUGLAS FIR	REDSTEM CEANOTHUS	SHINEYLEAF CEANOTHUS
	DOUGLAS FIR	REDSTEM CEANOTHUS	SCOULER WILLOW	GROUSE WHORTLEBERRY	MOUNTAIN ASH
	SERVICEBERRY		WESTERN RED CEDAR	RED ELDERBERRY	THINLEAF HUCKLEBERRY
	SHINEYLEAF CEANOTHUS		SHINEYLEAF CEANOTHU	ROSE	REDSTEM
			PACHISTIMA		PACHISTIMA
OTHER	SEDGE	CANADA BLUEGRASS	CANADA BLUEGRASS,	COLUMBIA BROME	SEDGE
	ARBOREAL LICHENS	SEDGES	COLUMBIA BROME	CANADA BROME	WESTERN NEEDLEGRASS
		IDAHO GOLDTHREAD	ORCHARDGRASS	CANADA BLUEGRASS	WOLF'S TRISETUM

	GREEN FALSE HELLEBORE	SEDGES	GREEN FALSE HELLEBORE	TIMOTHY
	COMMON BEARGRASS		WESTERN SHOWY ASTER	CANADA BROME
			COMMON BEARGRASS	

[1] Based on 57 rumen samples and 28 feeding site examinations.
[2] Based on 9 feeding site examinations
[3] Based on 37 rumen samples and 49 feeding site examinations
[4] Based on 23 rumen samples and 34 feeding site examinations for both years.
[5] Based on 22 rumen samples and 24 feeding site examinations.

Mountain dandelion

Indian paintbrush

American vetch

Strawberry

Sticky geranium Creeping barberry

Figure 8-1. Six flowering herbs that are palatable to elk. *See centrefold for this image in colour.*

Other studies indicate the principal browse species for elk in the Lochsa drainage are redstem ceanothus, Scouler willow, mountain maple, and serviceberry (Leege 1969, Asherin 1973). Asherin (1973) found that although redstem ceanothus was slightly lower in crude protein content than serviceberry, carbohydrate concentration in redstem was highest and elk selected redstem over other woody species. Asherin concluded that elk selected ceanothus because of high sugar content. Willows were highest in phosphorus and fats but was used most on recently burned sites and during periods of deep snows.

Distribution and Movements

There is a commonly held opinion by modern-day wildlife biologists to the effect that there is nothing like a sample of radio-collared animals to destroy older perceptions of what goes on. However, the perception that there was movement from the Fish Creek drainage on the Lochsa side over to the Weitas Creek drainage on the North Fork side, reported in 1936, was indeed substantiated with collared elk thirty-odd years later (Figure 8-2). Still, miscommunication and comments attributable to differing definitions and terminology resulted in confusion about movement patterns of elk in the Lochsa.

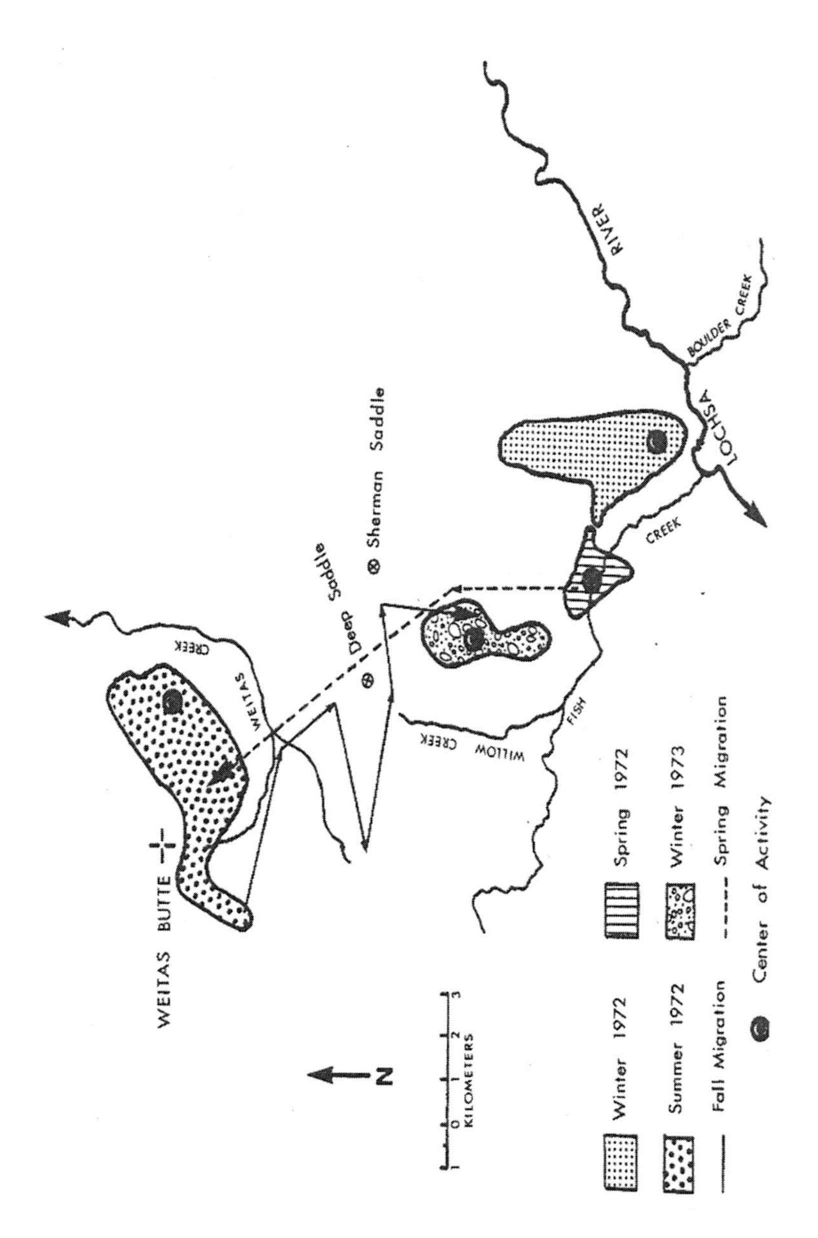

Figure 8-2. Movements from the Fish Creek winter range on the Lochsa to Weitas Creek on the North Fork of the Clearwater River (Hash 1973).

The USFS had decided to concentrate their winter surveys in the Weitas drainage during the winter of 1936-37, after a joint meeting of the Bitterroot, Lolo, Nez Perce, and Clearwater Forests in September concluded that:

> "inasmuch as the game situation in the Weitas drainage of the Musselshell District of the Clearwater represented the greatest problem from the standpoint of game management, studies on the Clearwater should be confined to this region."

At this time, the Lochsa drainage within the Clearwater National Forest was in the Selway Game Preserve. In 1935 and 1936, IDFG had opened the Preserve to elk hunting from September 15-30, inclusive, for the purpose of "relieving the upper Selway of over-stocking by removing 400 elk from that area." The season was controversial, with considerable hunting along the Preserve boundary during the regular elk season in the Weitas, which was thought to drive elk further into the preserve on the Lochsa. The 1936 report concluded that:

> "there was some ground for believing that the opening of the Selway Game Preserve to hunting last September may have caused a heavy drift from the Preserve into the Weitas. Ranger Jost reports seeing a band of 75 elk at the head of Bald Mountain Creek in the latter part of September, 1936."

In addition to being bounded by the Selway Game Preserve on the South, the Weitas was bounded by the Clearwater Game Preserve on the north. As a result, it was a major hunting area, part of the reason for the intense interest in it by USFS.

Norberg and Trout (1957) felt that the large wintering population of elk in the Lochsa did not have ample summer range because of the topography, and therefore many summered on the North Fork of the Clearwater. Movement between Pete King Creek and Lolo Creek further west was also suggested, and their map pointed to movement between the Weitas Creek and Fish Creek drainages north on the Lochsa. It remained for Judd (1971), McLean (1972), and Hash (1973) to confirm movement to Lolo Creek and Weitas Creek with radio-collared elk.

One radio-collared elk summered in the Sherman Creek-Bald Mountain Creek area, then moved into the head of the Weitas Creek drainage on Little Weitas Creek during the fall rut. Its winter range was along the Lochsa River in the vicinity of Sherman Creek. Another collared cow, trapped near Little Weitas Butte, summered on Little Weitas Creek and wintered on the Lochsa between Fish Creek and Sherman Creek. One radio-collared elk trapped on Pete King Creek summered in that drainage, then moved into the Lolo Creek drainage further west along the Middle Fork of the Clearwater to winter. Another cow trapped in the head of Fish Creek in summer utilized a summer area in that drainage, then spent the rut on Eldorado Creek in the Lolo Creek drainage, then spent the late fall in Pete King Creek where it was killed the last weekend of the hunting season.

Judd (1971) and McClean (1972) also observed elk that had been marked in the lower Selway River were in the lower Lochsa River in winter, spring and summer. Judd (1971) suspected that some intermingling with Montana elk along the Bitterroot divide also occurred. One collared animal was observed right on the divide. (M. Schlegel reported a long movement of marked elk from Fish Creek in April to 12 miles southeast of Superior, Montana in October).

McLean (1972) did not consider hunting pressure to have an effect on elk movement, unless an actual encounter with a hunter occurred. There were no general movements of elk related to whether a hunting season was in progress or not, instead being correlated more to onset of rut and subsequently to increasing snow depths at higher elevations. Generally, elk that used areas of high hunting pressure did not move appreciably. If not disturbed, elk may remain in close proximity to people during the hunting season. McLean (1972) observed one collared elk within 200 yards of two hunters and a vehicle, which remained in an area with hunter camps within a mile. Another animal spent the entire hunting season within a quarter mile of a road and one mile from an outfitter's camp. However, one collared animal, when contacted by a hunter, did move into a different drainage about an airline mile away from the area. Similar movements were reported by Irwin and Peek (1983) in the Coeur d'Alene River drainage, with elk moving outside of a currently occupied range to a different range following hunter contact. The pattern appears to be that elk do not move extensively during the hunting season after rut and before heavy snows, unless disturbed. Thus, the earlier observations of movement between game preserves and open areas which were interpreted as responses to changes in hunting pressure were subsequently found to be related primarily to other causes. However, a band of 75 elk observed moving from one drainage to another could indeed have been disturbed and thus were responding to direct human contact. Still, the observed movement of animals across major drainages often occurs during seasons when elk are not hunted.

The question remains as to whether hunting pressure early on could have precipitated these migrations to drainages outside of the Lochsa. The argument that animals disturbed by hunters (legally or illegally) would move a random distance outside of the occupied range in fright, sometimes across a major divide, and would then "remember" that area and subsequently include it within a seasonal home range. Such a one-time event would be very difficult to detect. However, there is no question but that elk are entirely capable of recognizing familiar terrain which is used only seasonally. Marked animals using the upper Lochsa which returned to the same winter range for more than one winter provide that evidence. One writer (Peek) observed elk moving over a 2,640 m pass in deep snow in early January during a period of intensive hunting pressure in the Gallatin River, Montana, to winter on range in the Madison River that had not been observed for a number of years.

Another argument is that elk first colonizing the head of a major drainage with a low saddle with suitable habitat between it and the adjacent drainage would simply use both drainages since both would be readily accessible and have suitable, similar habitat. Still another argument is that the Weitas Creek supplied a better forage base than the Lochsa side did.

Finally, animals dispersing from the home ranges of the maternal band as yearlings or two-year olds might establish new home ranges on suitable, similar habitat in adjacent drainages. Any of these hypotheses, in some combination or singly, could explain how movements across divides might be initiated and subsequently retained as a pattern of occupation.

The three studies initiated in 1969 and concluded in 1973 provided substantial information on movement patterns of elk in the Lochsa. Hash (1973) summarized the types of movement patterns the investigations identified, using radio-collared cows (Figure 8-3). First, elk wintering below the Black Canyon on the Lochsa, which is between Split Creek and Otterslide Creek, appear to migrate along the main river and mingle with elk above the canyon in summer. Elk that wintered above the Black Canyon either moved parallel to the river between summer and winter range or lateral to the river in side drainages. A high percentage of the elk wintering in Fish Creek or above underwent a seasonal migration, initially parallel to the river and then laterally up side drainages, with a few moving to the headwaters of the Lochsa. About 20% of the radio-collared elk did not move an appreciable distance between seasonal ranges, with changes in elevation being the most prominent aspect of their movement patterns.

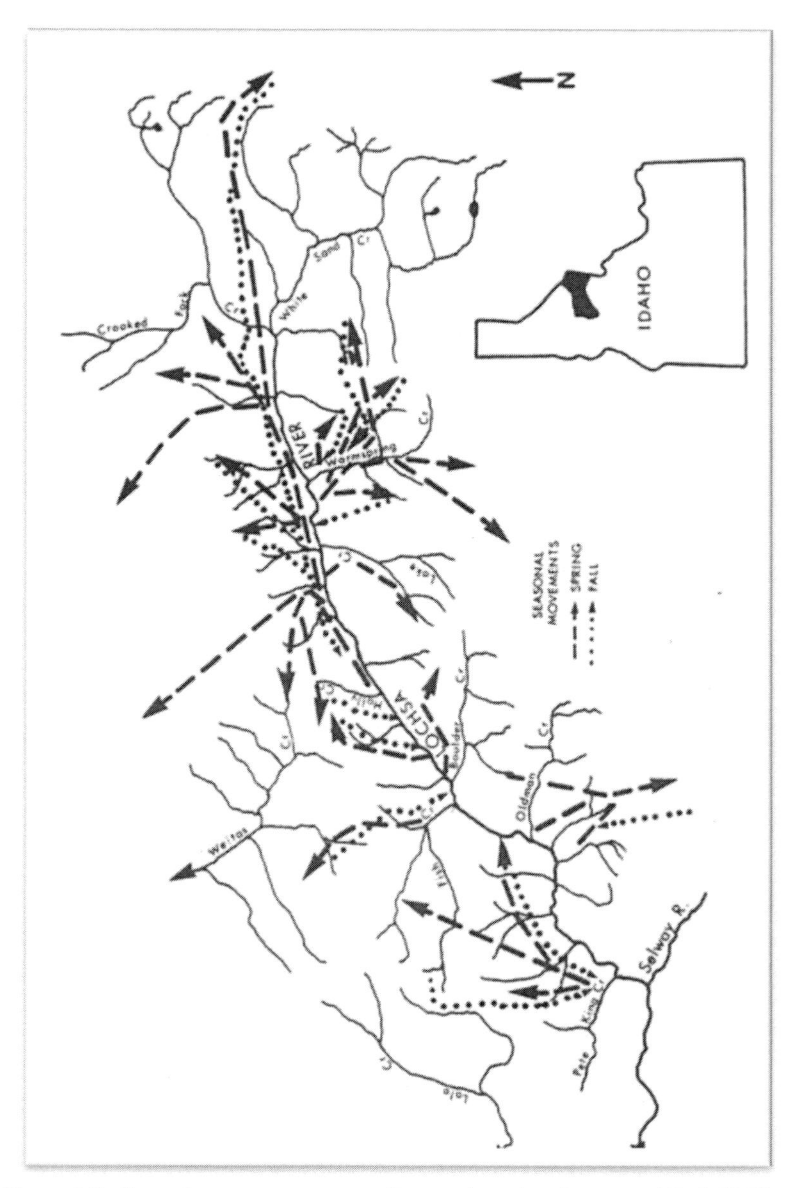

Figure 8-3. General movement patterns from major wintering areas (Hash 1973).

The short lateral-type movements which Norberg and Trout (1957) termed "lateral drifts" usually encompass a 3-8 km (2-5 mile) distance (Figure 8-4).

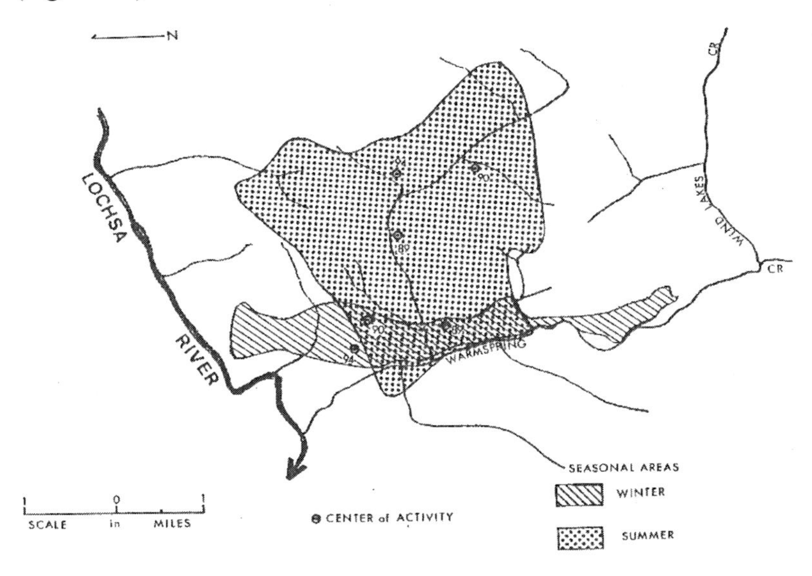

Figure 8-4. Local movements of three elk from March 1971 to January 1972 in the Warm Springs Creek drainage (Hash 1973).

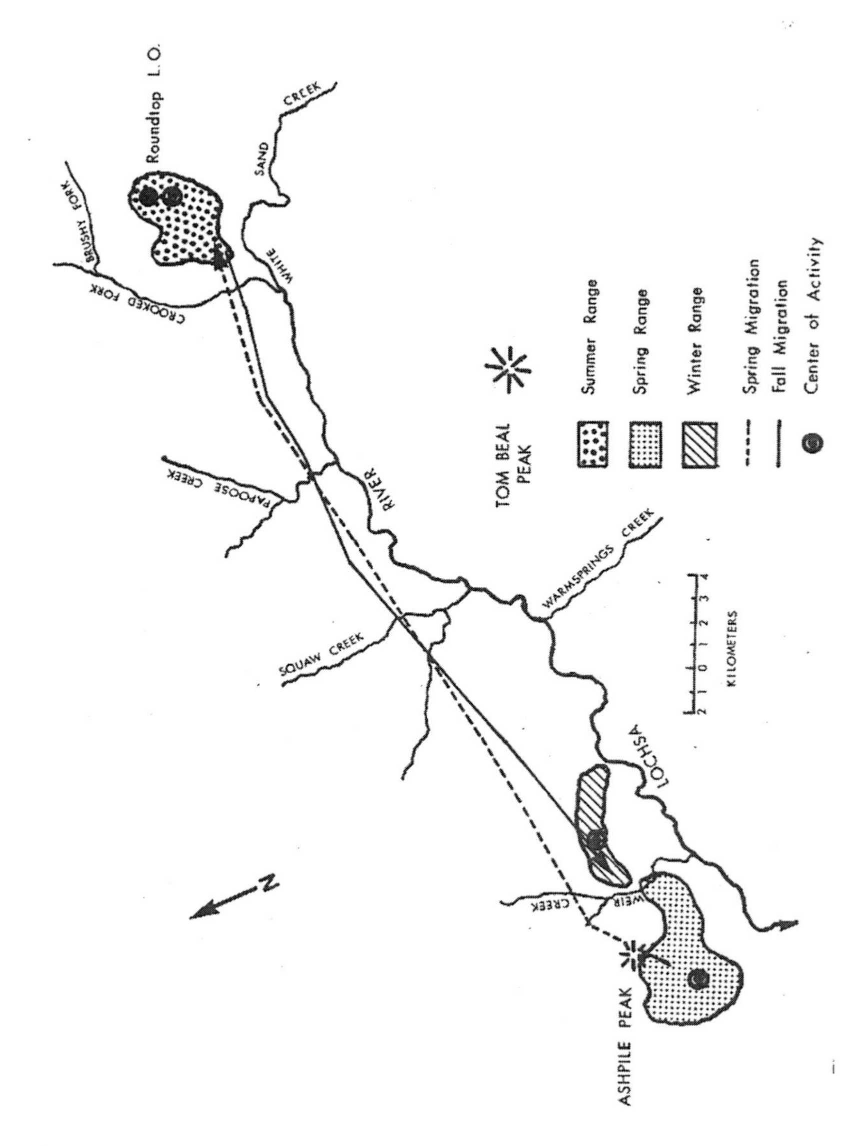

Figure 8-5. Long migration from Weir Creek to the head of the Lochsa (Hash 1973).

The long-lateral type movement of 5-10 miles involved a spring movement from low elevation wintering areas along the river up a major ridge between side drainages to a summer range. The long-parallel movement of 24- 56 km (15-35 miles) involved travel parallel to the river. Figure 8-5 shows an example of a long parallel migration from Weir Creek to Roundtop Lookout area (Hash 1973). One animal that migrated to the summit of the Bitterroot Range to several large wet meadows moved with the first light snowfall down the same route used in the spring.

These elk have high fidelity to the same ranges, except that the choice of winter range may be influenced by snow depth with animals moving further down river in deep snow years than in years of more shallow snow. Hash (1973) recorded one instrumented cow that used the same summer and winter areas for three consecutive years, separated by approximately 32 km (20 miles).

A number of factors probably influence movements, migrations, and habitat use patterns of elk, and there has been considerable interest in determining these for the Lochsa herd. Mention of deep snows forcing elk to drift to and concentrate at lower elevations occurs in the 1936 Forest Service report (Eddy and Rauma 1936). Williams (1962) reported studies involving extensive measurements of ambient temperature, wind, and snow depths and concluded that temperature and snow had greatest effect on elk movement and activity, and suggested further work be directed at snow depths and conditions.

The 1960-62 studies provided extensive information on microclimate. Air temperatures on ridges were consistently cooler at night and warmer during the day in shrub fields when compared to mature Douglas fir stands. Air temperatures on east- and west-facing slopes showed less variation than temperatures on ridges, all in shrub cover. There was considerable variation in maximum and minimum temperatures for four elevations (600-1,050 m, at 150 m intervals) and exposures. The range in temperatures from highest to lowest occurred at 600 m, 1,050 m, 750 m, and 900 m. Williams (1962) concluded that extreme temperatures were of little importance since his charts showed relatively rapid fluctuations. However, they may explain the presence of elk at different elevations when snow depths didn't over-ride the temperature differentials.

Air movement was generally greater in shrub fields than in mature timber, a phenomenon noticed by Ozoga (1973) in Michigan. Highest wind movement was recorded on brushy ridges, with decreased amounts on conifer-covered ridges, then west slopes, then east slopes. A daytime wind from valleys to mountains, and a nighttime wind downhill was dependent upon local topography for its intensity.

The pattern of snow depth was one of higher mid-winter depths in shrub fields and higher elevations than in conifers and at lower elevations. Steep slopes and southerly exposed aspects melted more quickly regardless of elevation. Conifer stands and northerly exposed slopes retained snow the longest.

Williams (1962) observed elk on southerly-exposed slopes with least snow in January 1962 when 61-76 cm had fallen in the Fish Creek area. As the snow crusted sufficiently to support elk, movement increased and bedding on ridgetops was commonly seen. Shrub fields received the major share of the use during the relatively severe winter. Elk did not seem to alter activity on rainy or snowy days, and usually more animals were observed during inclement weather than at other times. Shifts to conifer to bed as temperatures increased in late winter were noticed.

Subsequently, Leege and Hickey (1977) extended these surveys on the effects of snow on elk habitat use, using the Pete King drainage for study from 1967-1971. Elk were apparently sensitive to snow depths over 46 cm, and most observations occurred in areas where depths were less. Nine aerial surveys of elk distribution in relation to snow depth were made. The majority of elk were observed in old shrub fields. Again, use of ridgetops increased as snow depth increased, as did use of south-facing aspects. During the deeper snow portions of the winters, lower elevations received more use. Browse utilization did not change between elevations for the five years of study, except in timing of use which reflected distribution of animals and winter severity. Use of mature conifer cover to escape from the helicopter was noted, indicating elk were sensitive to the presence of the noisy aircraft. Use of mature conifer cover at the lowest elevation, when snow depths over 61 cm existed, also increased. This study occurred during a series of relatively mild winters, with only one winter (1968-69) being above the norm for the 1939-1974 period (Figure 2-2). Leege and Hickey (1977) recognized that many elk are found to winter commonly in over two feet of snow, suggesting that survival is related to their condition going into the severe winter period. The length of the winter period of deep snow also is probably important. Provision of adequate amounts of dense canopied old-growth conifer in proximity to productive shrub-fields was recommended. Accordingly, areas which commonly have less than 46 cm of snow (in Pete King Creek this would be south aspects to 1,230 m, west aspects to 1,050 m, east aspects to 990 m and north aspects to 870 m elevation) should receive high priority for habitat management, and some combination of mature conifer cover and productive shrub fields should be created and/or retained.

McLean (1972) related elk movements to snow depth, ambient temperature, and plant phenology. Increasing snow depths were thought to exert the strongest influence on initiating fall movement to winter range, who also noted that Norberg and Trout (1957) reported extended periods of cold weather also caused downward movement. On winter range snow depths impede movement if soft snows prevail, while elk may move up in elevation on warm winter days that cause the snow to pack and crust.

Plant development and temperature may be primarily influential in determining elk movements in spring and between habitats in summer. McLean (1972) reported the relationship observed in 1971 for elk in the Lochsa. General plant development appeared to stimulate movement to higher elevations, but calving activities may stop movement and obscure relationships. Use of separate spring range during calving, and separate range during fall rut was also noted for some individuals.

Mysterud (2013) summarized work on causes of migration which involved conventional thinking that snow depths promote downward movement to winter range and delayed plant production that causes plants at higher elevations to be more palatable later on promotes spring migration. However, Mysterud (2013) concluded that presence of predators can also influence migratory behavior as documented in southwestern Alberta (Hebblewhite and Merrill 2009, 2011) and in the Yellowstone (Middleton et al. 2013). Shifts in elk distribution away from areas where wolf predation is intensive may mean reductions in migratory behavior or shifts to either higher or lower elevations where wolves are not present.

The radio-collared elk data suggest that some individuals undergo fairly long distance movements between distinct winter and summer ranges, that are more appropriately migrations (Fryxell and Sinclair 1988). These data contrast with a statement in the 1934-35 winter game study report of the Nez Perce-Bitterroot-Clearwater which stated:

> "True migrations on the Selway and Lochsa are nonexistent. Due to the character of the topography game movement is reduced to minimum; the game in general moving to a lower elevation during the winter, with a contracting of the game range from the upper reaches of all large drainages."

Obviously, the older studies did not consider the movements they observed which would now be termed migrations to be more than localized shifts. However, if a migration is defined as a movement away from one seasonal home range to another distinctly separate home range used in another season, with a return, to distinguish it from emigration and dispersal which indicated movement without return, then some of the observed movement patterns are true migrations.

Home Ranges

Home ranges of the radio-collared elk were estimated by determining a geometric center of activity, and then the standard diameter describing the area encompassing 68.3% of all relocations about that center of activity, following Hayne (1949) and Harrison (1958). Data for the radio-collared cows are presented in Table 8-3. Problems in using the technique are typically associated with sample sizes, but no statistical trends in size of mean standard diameters related to number of relocations were apparent. Extensive variation in size of home ranges between individuals, seasons, and years, occurred. Generally, home ranges were larger in 1972 than 1971, which may reflect poorer growing conditions in 1971, or simple sampling error. The 1972 data included a few cows with extremely large home ranges that were not evident in the 1971 samples. There was no correlation between size of winter range and winter severity because the first two winters were mild and the last winter was intermediate between the other two. The bias is considered strong enough to preclude use of the information to represent changes in size of seasonal home range between years or seasons, and all data must be considered minimum estimates. However, the migration patterns are quite likely accurately portrayed. The expense and logistics of obtaining adequate sample sizes to determine seasonal home ranges in the steep relatively inaccessible mountain country are well illustrated by these data.

Table 8-3. Standard diameters of elk seasonal home ranges, 1971-1972 (McLean 1972, Hash 1973).

SEASON	YEAR	SIZE OF HOME RANGE				NUMBER OF ELK.
		MEAN acres	RANGE acres	MEAN km^{-1}	RANGE km^{-1}	
Winter	1971	8.32	365-1427	3.4	1.5-5.8	14
	1972	1958	448-17.13	7.98	1.81-44.37	11
Spring	1971	947	628-1408	3.83	2.53-5.70	8
	1972	4166	1114-16115	16.56	4.51-65.22	12
Summer	1971	1171	704-2234	4.74	2.85-9.03	20
	1972	2496	394-6451	10.10	3.21-26.11	10
fall	1971	1510	928-2534	6.11	8.76-10.25	4
	1972	4262	493-12544	17.24	1.99-50.76	10

Habitat Use

McLean's (1972) study provided further insights into habitat use of elk in the Lochsa (Figure 8-6a). Again, high use of mature timber during the deepest snow period was seen for all elk observed. However, three radiocollared elk made more use of regrowth timber during the heavy snow month in Pete King Creek where Leege and Hickey (1977) also observed this pattern (Figure 8-6b). These relocations were of resident animals using conifer regrowth near open shrubfields, predominantly the controlled burns. Large shrub fields in the multiple-burned areas characteristic of the lower and middle portions of the Lochsa received the highest percentage of observations in late winter and early spring. Increasing use of timbered areas occurred from spring to a peak in early fall.

Neither Leege and Hickey (1977) nor McLean (1972) observed much use in clear-cut areas during summer, quite different from observations elsewhere in northern Idaho (Irwin and Peek 1983, Alldredge et al. 2002). Leege and Hickey (1977) considered the extensive road system and resultant human activity contributed to the lack of use of logged areas with suitable forage. This pattern has been observed consistently elsewhere (Christensen et al. 1991), prompting recommendations to shorten human use and institute management of road traffic in areas where elk are desired to be present.

Bulls were observed early on to winter at higher elevations or at least apart from cows and calves, but the information from direct observation and radiotelemetry studies in the 1960s and 1970s was primarily obtained from the female segment of the elk population. Schlegel (1972) reported that 344 elk were trapped, ear-tagged and released between 1965 and 1968 on the Coolwater Ridge study area, of which 48 were returned by hunters and 1 tag found at a cougar kill site (Figure 8-7). Tag returns the year after capture for 1966, 1967, and 1968 were 7.8%, 5.7%, and 5.8%, respectively. The three-year average was 6.1%. If the tagged animals killed by hunters were reported, these percentages indicate the percent of the elk herd is harvested. While some successful hunters will not report a marked animal, Idaho hunters are probably similar to hunters in other states and the percent of cooperative hunters should be comparable. If this is the case, then the harvest rate on the Coolwater Ridge is low and not sufficient to have an effect on the population. The probable explanation is that these elk exist in areas of poor access, the terrain is very steep and the brushfields provide excellent escape cover, making hunting difficult.These elk were examples of local movements without evidence of migratory behavior.

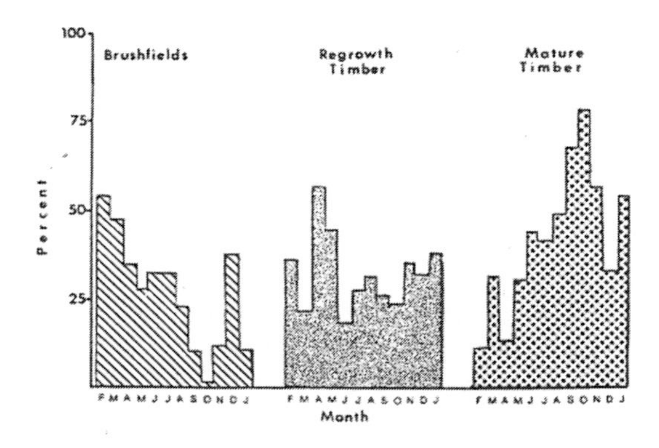

Figure 8-6a. Use of three cover types by month after McLean 1972)

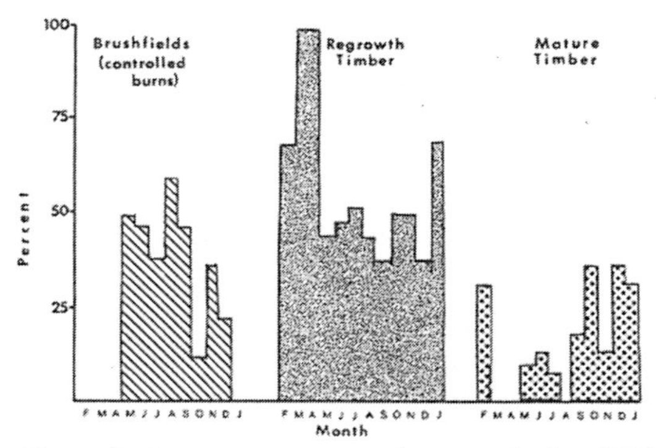

Figure 8-6b. Use of three cover types by month in 1971 after Leege and Hickey (1973)

Concerns over bull harvest intensity and lack of knowledge of ecological relationships of mature bulls prompted more investigations in the mid-1980s. Unsworth et al. (1991) summarized the information from the 1985-1990 period for the fall hunting seasons (Table 8-4). The tendency to use mature closed timber was evident for all animals, but bulls taken during hunting used higher percentages of closed canopy forest adjacent to clear-cuts in roaded portions of the Lochsa than did bulls not killed or cows (no hunting of cows during the period of study). The information tended to corroborate the earlier investigations that suggested that human disturbance had an effect on elk distribution and habitat use, in this case by selectively eliminating bulls closest to access in managed forests. The more recent work demonstrated the shift to denser cover in fall by both sexes which appears to be a general pattern of habitat use by elk throughout the northwest (Unsworth 1991).

Table 8-4. Autumn use of cover types from 1985-1991. A + or - indicates a statistically significant difference from available percentage of each type (Unsworth et al. 1998).

		-------	% USE	------	-------
COVER TYPE		AUG.	SEPT.	OCT.	NOV.
Closed Timber	Cow	36	64+	51+	11
	Surviving bull	55+	58+	53+	23
	Harvested bull	63+	52+	52+	33
Open Timber	Cow	44	29	40	28
	Surviving bull	26-	26-	32-	31
	Harvested bull	21-	31-	24-	40
Shrubfield	Cow	16	6-	7-	28
	Surviving bull	15	12-	14-	32+
	Harvested bull	11-	9-	12-	27
Harvested timber	Cow	5-	2-	2-	33
	Surviving bull	4-	4-	2-	13
	Harvested bull	5-	7-	4-	0
Other	Cow	0-	0-	0.6-	0
	Surviving bull	0-	2-	0.4-	1
	Harvested bull	0.6	2	1	0

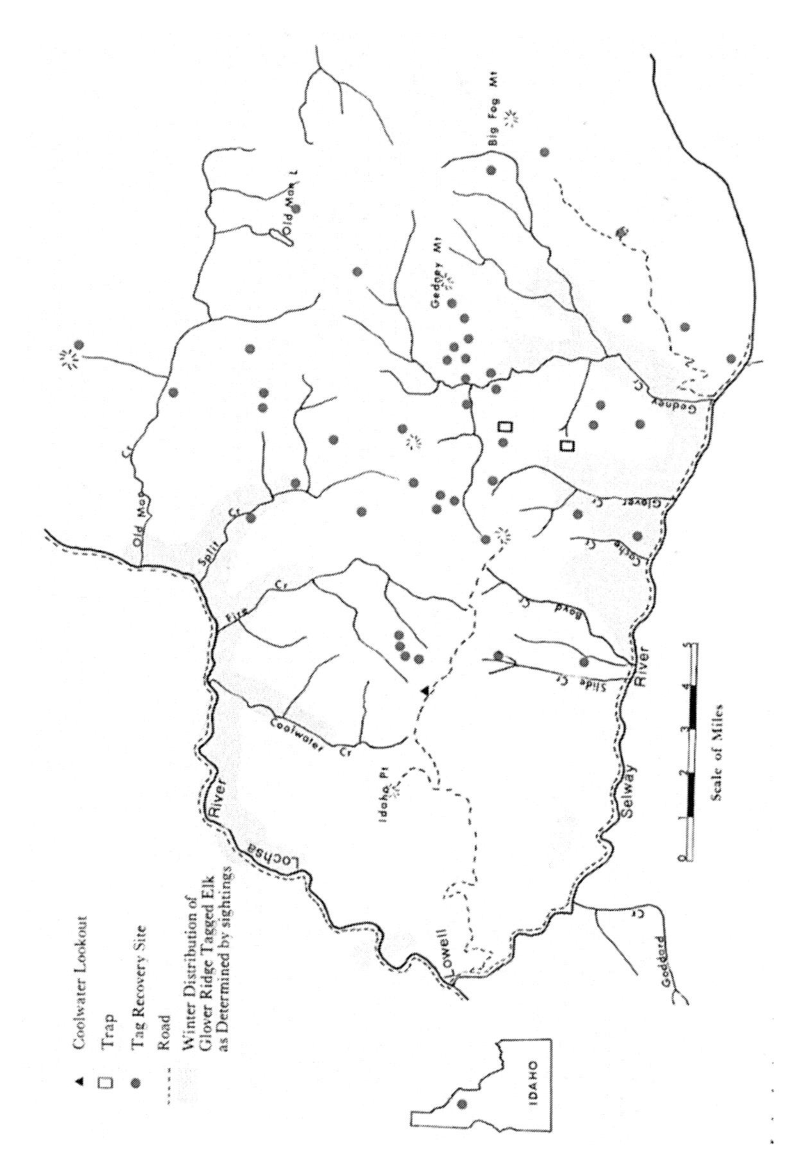

Figure 8-7. Locations of harvested elk that were marked near Glover Ridge in the Coolwater Ridge study area (Schlegel 1972). These elk did not move much from the area they were marked. Red dots show the area they were killed and the two squares show the trap sites.

Although winter distribution patterns are undoubtedly reflected by winter severity, the distribution of elk along the Lochsa drainage has been remarkably similar over a period of years (Table 8-5). An average of 40% of the elk observed over 11 years of flights from 1957 to 1982 were in the lower Lochsa River from Otterslide Creek to Lowell. Fish Creek has consistently wintered about a fourth of the elk population. The Boulder-Lake Creek segment of the river has consistently wintered about 28% of the population, while 12% has wintered above Lake Creek (Table 8-5). The 1957 distributions reported from the Norberg and Trout (1957) surveys are at most variation from the 1969-1982 series of surveys, with few elk wintering on the lower reaches of the drainage and more wintering above Lake Creek. This may reflect more rapid succession of the shrub fields to conifer stands in the upper reaches of the drainage, where elk have shifted down river since. There is no correlation between winter severity index and the proportions of elk on the upper and lower reaches of the drainage.

Also, there was no correlation between winter severity index and number of elk seen, suggesting that visibility bias related to proportion of conifer cover within the various stretches of the drainage has not appreciably influenced the number of elk observed. More likely, weather and snow conditions during the surveys influence numbers of elk seen more than general winter conditions, as Williams' (1962) descriptions of elk habitat use relative to microclimate and snow suggest.

The 1969-1982 period had two severe winters (1971-72, 1978-79) and the remainder were relatively moderate to mild. The possibility exists that winters were not difficult or severe enough to demonstrate the variation in winter distributions which might occur if more severe winters were included. Also, the period is characterized by low to slowly increasing elk populations, which may also affect winter distribution. Nevertheless, the available information suggests that, while a series of different migration patterns prevail, the distribution of the wintering population is relatively predictable. Thus, shifts in distribution probably are more related to elevational and local shifts within the broader regions of the drainage, than between areas.

Table 8-5. Winter elk distributions 1956-1982. Idaho Department of Fish and Game data.

	MILES^2	MILES^2	MILES^2												
	WINTER	BURNED	MULTIPLE	-----------PERCENTAGE OF ELK OBSERVED IN EACH AREA----------											
AREA	RANGE		BURNED	1956	1969	1974	1975	1976	1977	1978	1979	1980	1981	1982	MEAN
Lowell-. Otterslide Ck	40	32.9	12.5	21	48	38	52	44	38	33	49	41	37	37	40
Fish Ck - Sardine Ck.	11.9	11.9	8.7	24	21	23	19	19	18	18	16	21	24	22	21
Boulder- Lake Ck.	33.6	28	25.5	17	25	29	26	29	32	33	28	26	35	30	28
Above Lake Ck.	31.6	15.3	0.4	38	11	7	4	8	12	16	7	13	4	11	12
Total seen				3567	2492	1257	1166	1469	1280	1376	1689	1683	1384	2323	1790

Differences in distribution and habitat use by cows and bulls were reported for the Lochsa elk population using 121 radio-collared elk (101 bulls, 20 cows, Unsworth et al. 1998). Winter selection patterns of cows showed more use of open shrub-dominated habitats and less open timber types than bulls. This was most evident in December-February while cows used timber types more in March, similar to bulls. Spring-summer-fall ranges showed similar use patterns for the sexes, although bulls tended to use lower slopes and stream bottoms more than cows. Elk in roaded areas showed preferences for forested areas as compared to elk in unroaded areas that used open timber. Older bulls also wintered in deep snow areas at higher elevations. Elk of both sexes used greening vegetation in spring except use of timbered types associated with recreational use and logging activity was most used by elk associated with human disturbance. Yearling bulls used different habitats in summer and fall than older bulls, more closely resembling use patterns of cows.

Green and Talbert (1991) initiated evaluations of distribution of elk pellet groups, key shrubs and soils on the north side of the Selway River within 5 miles of its confluence with the Lochsa River. This area was divided into three elevation zones (2100-2500 feet, 3000-3500 feet, and 4000-4500 feet), north and south aspects, and five burn years to provide for a quantitative analysis of the data. A total of 64 sites were examined for pellet group densities, densities of Scouler willow and redstem ceanothus, and soil descriptions. This area is west of the Selway-Bitterroot Wilderness boundary and was subject to a series of prescribed fires in 1985 and 1989. These were hot fall burns that were more typical of the natural fire regime than the cooler spring burns which do not affect soils Hooker (1972). The area had been burned in the 1910-1934 wildfires and there was an unburned designation for stands that had not been burned since 1900, for the 5 burn years.

As expected, elk pellet group densities were highest on the recently burned plots (Table 8-6). The 1985 burn contained the highest densities, associated with the high densities of redstem ceanothus and Scouler willow. South-facing slopes were used more than north faces and the higher elevations were used more than the lowest elevation. While pellet group surveys have been criticized for not reflecting actual use patterns, the size of the area to be examined can reduce bias if the sample sizes are large, and personnel are diligent about the field work.

Table 8-6. Distribution of elk pellet groups by elevation, aspect, and burn age, summer 1991, lower Selway River (Peek et al. 1992).

BURN YEAR	ELEVATION Feet (meters)	ASPECT	MEAN $/10m^{-1}$	SD	ELEVATION MEAN/ $10m^{-1}$
UNBURN	2000-2500 (600-750)	S	0.68	0.75	0.63
UNBURN	2000-2500 (600-750)	N	0.45	0.65	
UNBURN	3000-3500 (900-1050)	S	0.53	0.83	1.03
UNBURN	3000-3500 (900-1050)	N	0.28	0.52	
1910-1934	2000-2500 (600-750)	S	0.23	0.53	
1910-1934	2000-2500 (600-750)	N	0.05	0.22	
1910-1934	3000-3500 (900-1050)	S	0.27	0.52	
1910-1934	3000-3500 (900-1050)	N	0.73	0.99	
1910-1934	4000-4500 (1200-1375)	S	0.18	0.47	0.91
1910-1934	4000-4500 (1200-1375)	N	0.4	0.67	
1985	3000-3500 (900-1050)	S	2.83	1.56	
1985	3000-3500 (900-1050)	N	1.97	2.25	
1985	4000-4500 (1200-1375)	S	2.12	1.5	
1985	4000-4500 (1200-1375)	N	0.92	1.28	
1989	2000-2500 (600-750)	S	0.75	0.93	
1989	3000-3500 (900-1050)	N	0.47	0.72	
1989	3000-3500 (900-1050)	S	1.15	1.25	
1989	4000-4500 (1200-1375)	N	0.83	0.89	
1989	4000-4500 (1200-1375)	S	1.00	0.9	

Green and Talbert (1991) concluded that soil organic matter levels are decreased by burning, corroborated by other studies in the region, and do not recover to levels of mature forests at 25 years. Mean natural fire frequency of 50-200 years may be more typical in these grand fir and western red cedar habitat types in this study area. Soil degradation has likely occurred on the Selway sites due to multiple fires. A volcanic ash layer found on many sites in the region, was generally absent in the plots that were examined.

It remained for the work with conditioned elk at the Starkey Experimental Forest in northeastern Oregon to clarify the role of conifer cover for elk. A number of earlier reports used the assumption that closed canopy mature forest of trees over 40 feet tall with canopy closure of 70 per cent was needed (Thomas 1979). However, the research using conditioned elk showed that animals that were placed in that kind of cover for the winter actually lost weight while individuals put in open areas did much better (Cook et al. 1998). This does not mean that closed canopy forest will not be used during inclement weather and other situations for a period of time, however.

Since the Lochsa elk population expanded during a period when the fires had virtually eliminated conifer overstories, and elk in the more open country such as Yellowstone National Park and much of southwestern Montana can be highly productive and exist at high densities, these findings were not unexpected. Elk are known to winter at the highest elevations where no conifer cover is available but high quality forage is present. They will bed on the lee sides of ridges out of the wind and feed on sites that the prevailing wind has bared off. This can be observed on places in the James McClure Jerry Peak Wilderness in east central Idaho and on Antone Peak in the Snowcrest Range in southwestern Montana. Their cousins the red deer of Scotland, have little or no conifer cover on most of their range. Snows in the Lochsa are typically denser, making them resistant to winds, however. Nevertheless, the lack of cover in the Lochsa was outweighed by the large quantities of forage that was available during most winters following the big fires.

Habitat use patterns are known to change for a variety of reasons as the information suggests. At low densities, individuals should select the most preferred habitats since forage and cover should be readily accessible (Perez-Barberia et al. 2013). As populations increase and the availability of resources for a given individual declines, some individuals may have to use less preferred habitats or physical condition of others may be affected. This assumes that habitats provide a range of forage and cover conditions that elk have opportunities to select from. Also, predators, especially wolves, are known to influence habitat use and distribution, causing elk to occur in smaller groups and to use cover including coniferous forests more than without the presence of predators (Boyce 2018). The available information was mostly obtained prior to the presence of wolves. Thus, the current situation in the Lochsa may not be necessarily be reflected in the information that was obtained in earlier years when habitat conditions were better for the elk, population densities were higher, and wolves were not present.

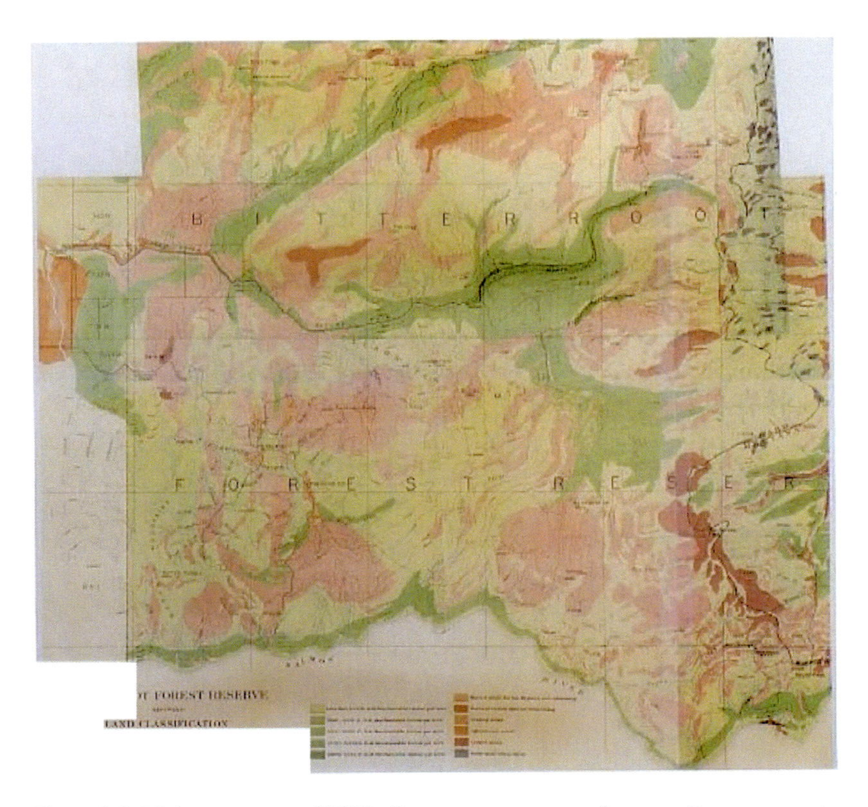

Figure 3-2. Lieberg surveys of 1899. Green areas represent forest, yellow represents regenerating forest, and red represents shrublands and other areas that are nonforested. The uppermost green is the cedar stands along the Lochsa River. The middle green is the Selway River with the darkest green being Moose Creek. The thin lowest green is the Salmon River.

Figure 7-3. Trail in Dutch Creek showing the predominance of shrubs. Taken in the 1950s by George Hatley.

Mountain dandelion Indian paintbrush

American vetch Strawberry

Sticky geranium Creeping barberry

Figure 8-1. Six flowering herbs that are palatable to elk.

CHAPTER NINE

FOREST SUCCESSION

Plant community changes are, relative to most issues that humans deal with, slow processes. However, the process that follows disturbance, whether it be human caused or naturally caused, is predictable to some extent. The north Idaho forests that include those in the Lochsa that are severely burned will initially develop a plant community that primarily consists of forbs and grasses. After a few years, shrubs will dominate the burned area, and then the coniferous species will begin to appear. Often mixes of species such as western larch, white pine, lodgepole pine, and Douglas fir will dominate the sites for a few centuries or so. Eventually the so-called climax species which can include Ponderosa pine on the driest slopes to western red cedar on the wet slopes and riparian zones, to mountain hemlock, subalpine fir and Engelmann spruce on the higher elevations, will occupy the sites. This process may not happen in one's lifetime, and it can be disrupted by natural and human-caused events.

A sequence of forest development following a fire is portrayed in Figure 9-1. This sequence is for a grand fir/ pachistima stand since that forest habitat type is common in the Clearwater River drainage. Douglas fir is typically the species that occupies the site following fire in this type. Still, one may drive up the Lochsa River today and see shrub fields that are over 80 years old. The reason for retention of these shrub fields is that the conifer trees that would in other circumstances invade and dominate the site have been browsed back by elk. Hobbs (1996) concluded from a review that ungulates are important factors in altering successional processes and switching of plant communities from one state to another. The Lochsa is an excellent example of these processes. On the more north-facing slopes, conifer succession has advanced to where dense pole stage stands have replaced the shrub communities, but was still slowed in some areas by elk browsing from what it would have been without their presence.

Wildfire has been prevalent since the retreat of the continental glacier system in this general area for at least 12,000 years as determined from investigations of a bog near Lost Trail Pass in the Bitterroot Mountains south of the Lochsa (Mehringer et al 1977). Fire scars on old trees and stumps plus written records suggest 35 major fire episodes from 1540-1940 in the inland northwest, including the Lochsa (Barrett et al 1997). Ponderosa pine, lodgepole pine, Douglas fir, and western larch forests were the most extensive types in the region, all known to be adapted in different ways to fire.

The thick, fire-resistent bark of mature ponderosa pine, western larch, and Douglas fir trees tends to minimize mortality from the typical ground fire that prevailed. Lodgepole pine cones open as heat from the fires causes the resins to liquefy, allowing the seeds embedded in the cones to drop and subsequently germinate. Fires prior to 1900 were of mixed severity, more so than after that time, when stand replacement fires became more prevalent. Fire intervals of 20 years in ponderosa pine forests, 52-year intervals in Douglas fir and western larch forests, and 112 years in lodgepole pine forests prior to 1900 were estimated by Arno (1980) and Barrett et al (1997). These forest types were common in the Lochsa, although grand fir, western red cedar, and spruce/fir forests were present and would represent the eventual composition of the forests if no disturbance occurred.

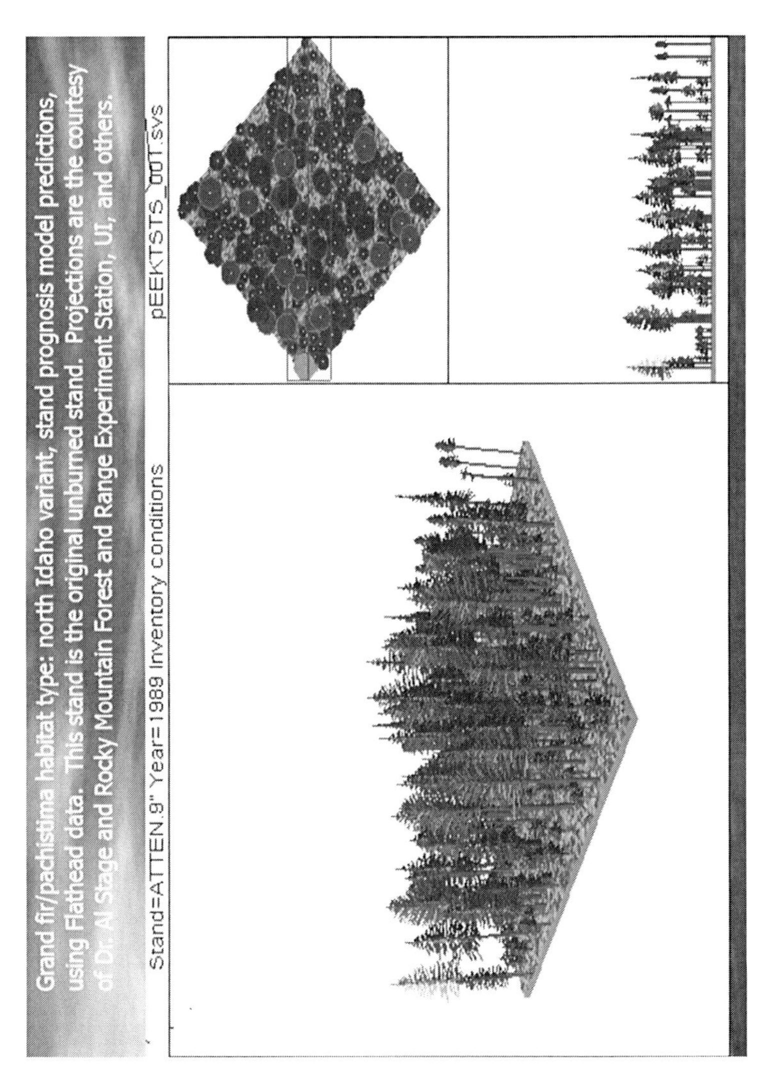

Figure 9-1. A sequence of plant succession following a fire in grand fir/boxwood habitat type derived from the stand prognosis model (Stage 1973, Moeur 1985, Ferguson and Carlson 1993). Essentially, the shrub understory that develops between 5 and 20 years following the fire will start to decline at 25 years and will be replaced by conifers at 40 years. This is without any browsing.

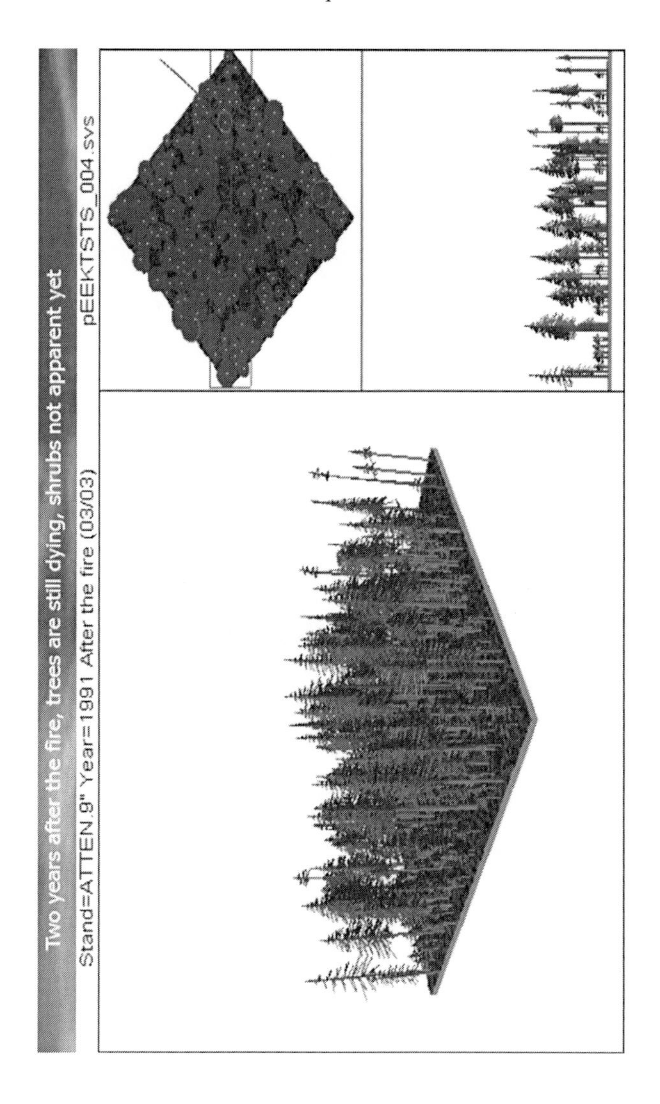
Two years after the fire, trees are still dying, shrubs not apparent yet
Stand=ATTEN.9" Year=1991 After the fire (03/03)
pEEKTSTS_004.svs

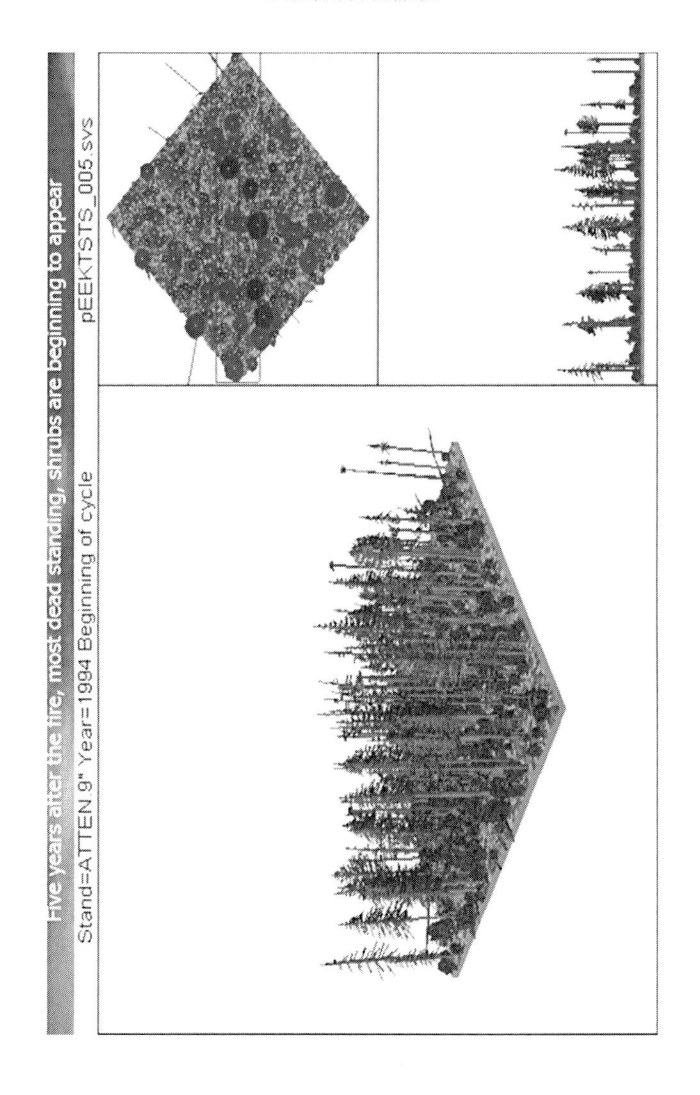
Five years after the fire, most dead standing, shrubs are beginning to appear
Stand=ATTEN.9" Year=1994 Beginning of cycle
pEEKTSTS_005.svs

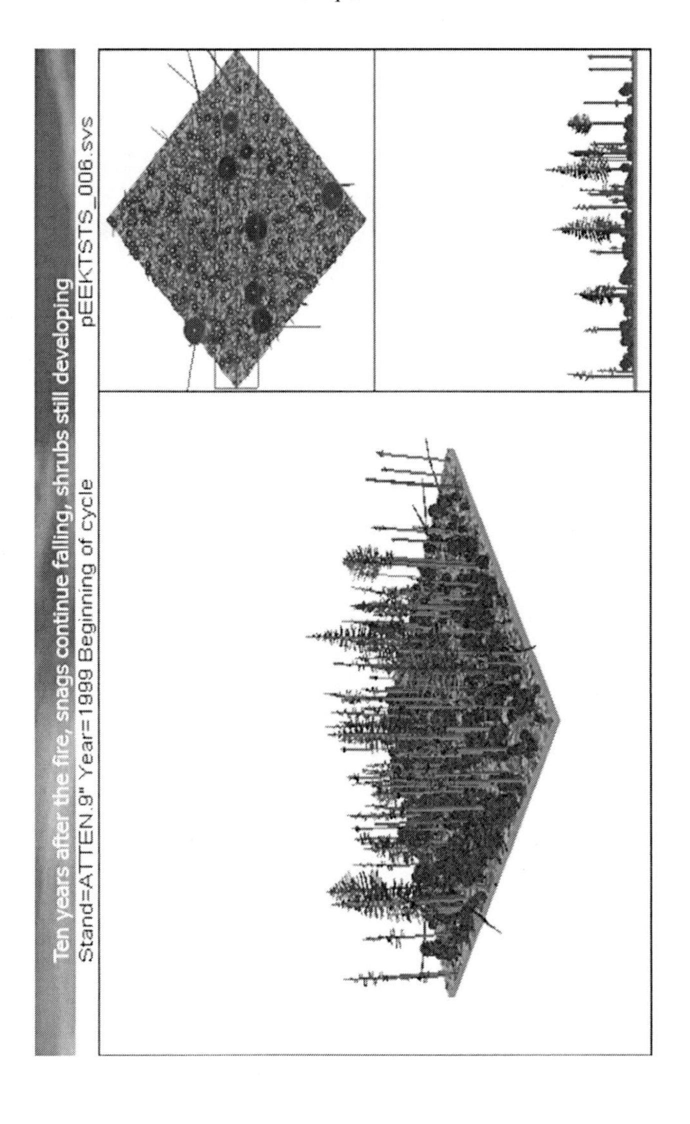
Ten years after the fire, snags continue falling, shrubs still developing
Stand=ATTEN.9" Year=1999 Beginning of cycle
pEEKTSTS_006.svs

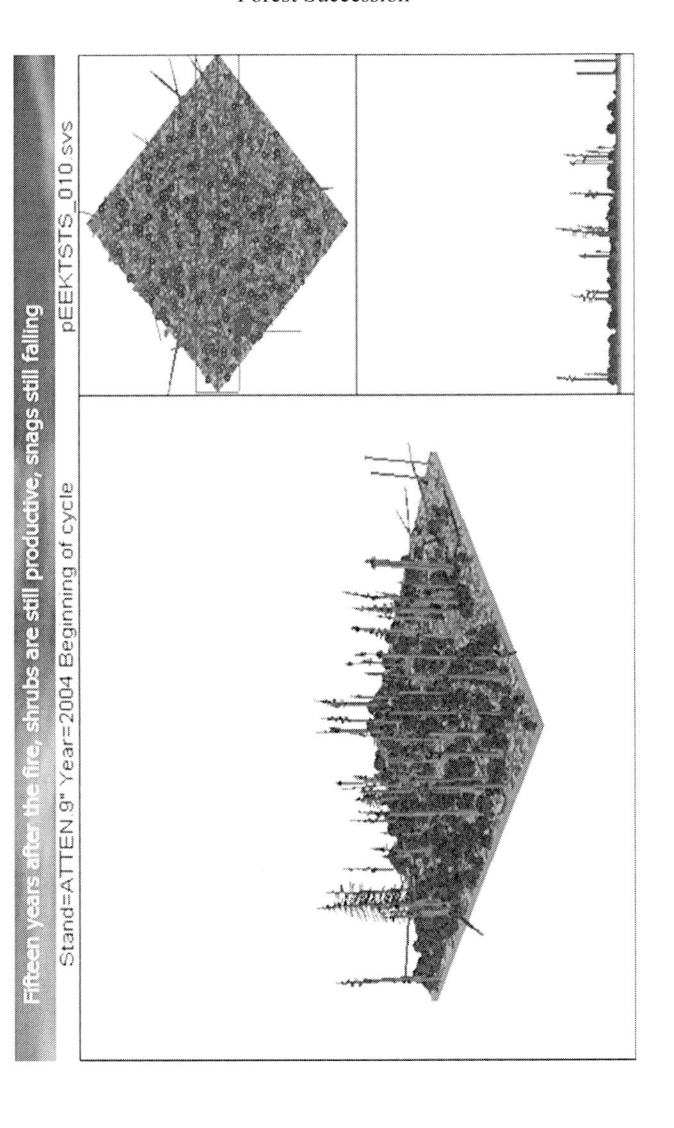
Fifteen years after the fire, shrubs are still productive, snags still falling
Stand=ATTEN.9" Year=2004 Beginning of cycle
pEEKTSTS_010.svs

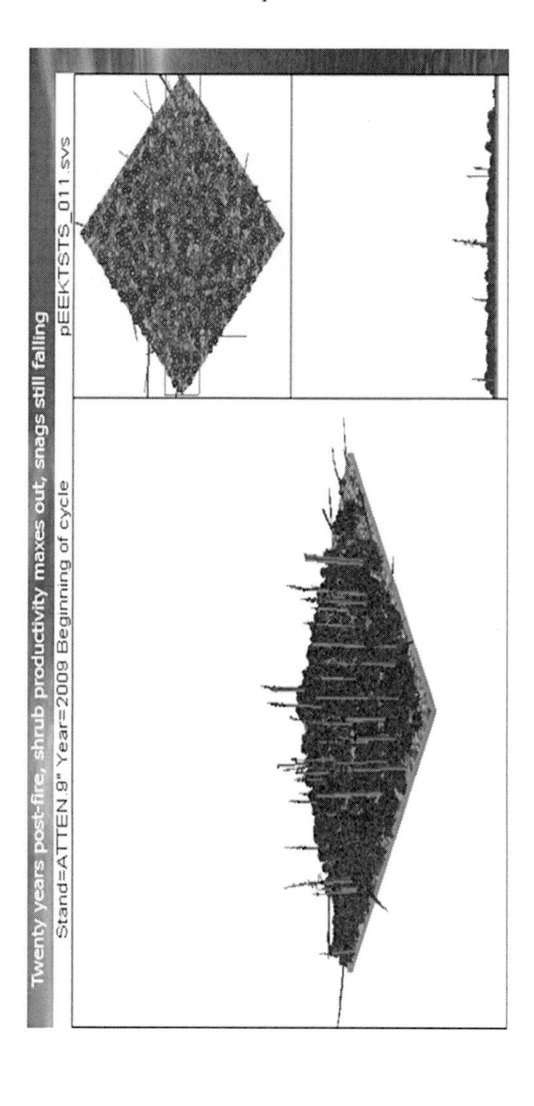
Twenty years post-fire, shrub productivity maxes out, snags still falling
Stand=ATTEN.9" Year=2009 Beginning of cycle
pEEKTSTS_011.svs

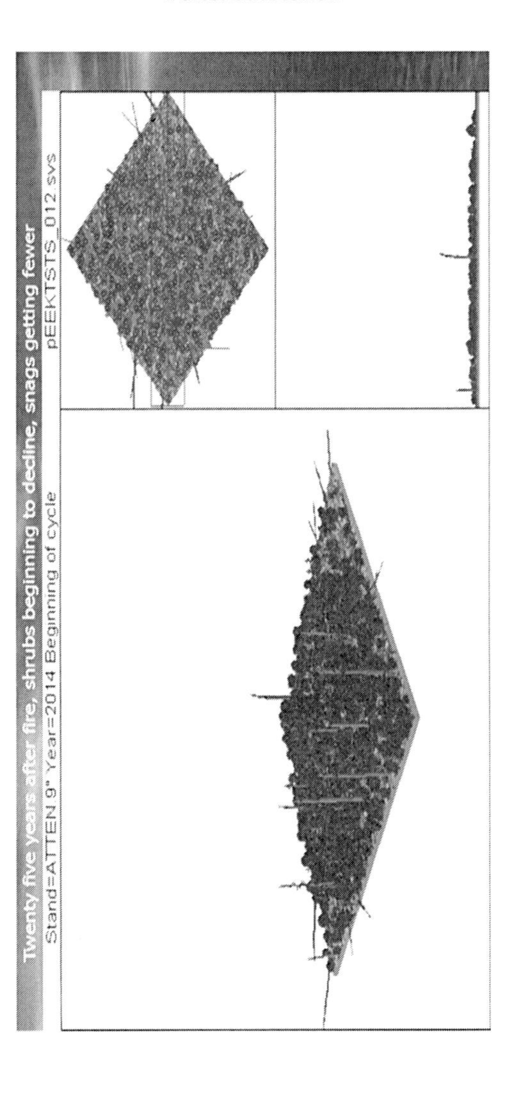
Twenty five years after fire, shrubs beginning to decline, snags getting fewer
Stand=ATTEN 9" Year=2014 Beginning of cycle
pEEKTSTS_012.svs

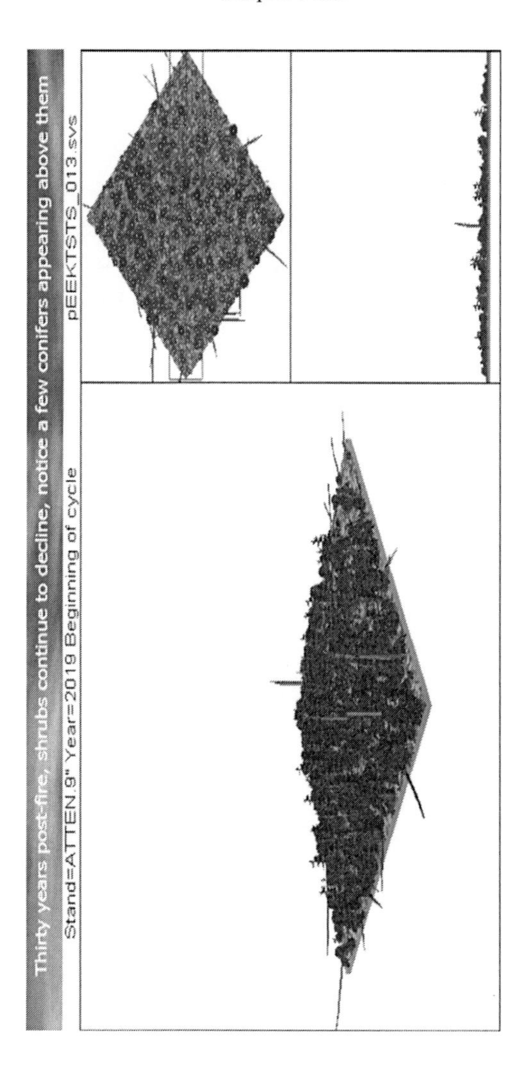
Thirty years post-fire, shrubs continue to decline, notice a few conifers appearing above them
Stand=ATTEN.9" Year=2019 Beginning of cycle
pEEKTSTS_013.svs

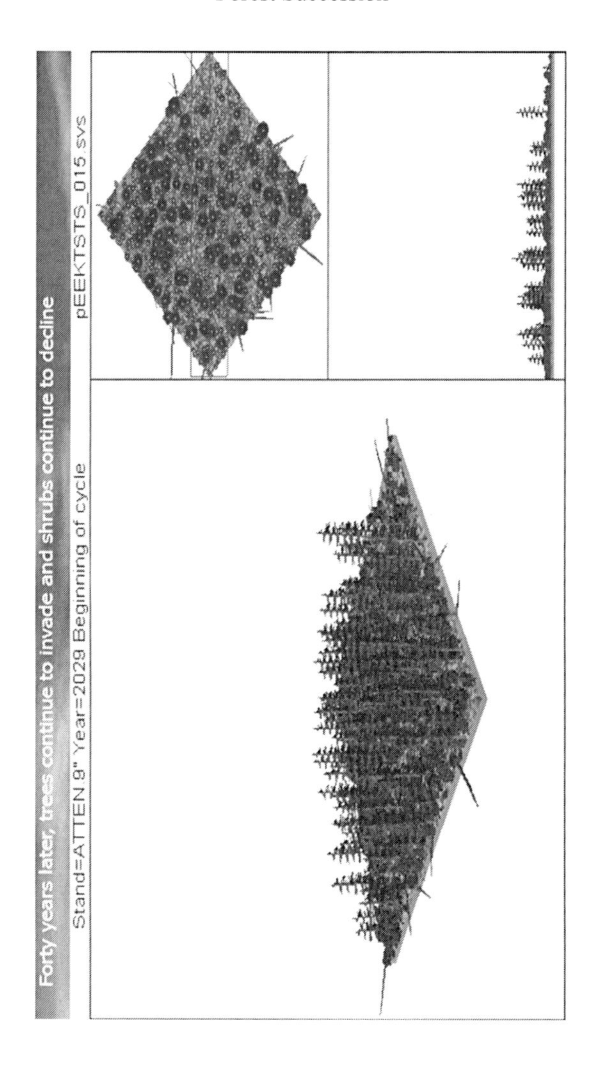
Forty years later, trees continue to invade and shrubs continue to decline
Stand=ATTEN 9" Year=2029 Beginning of cycle
pEEKTSTS_015.svs

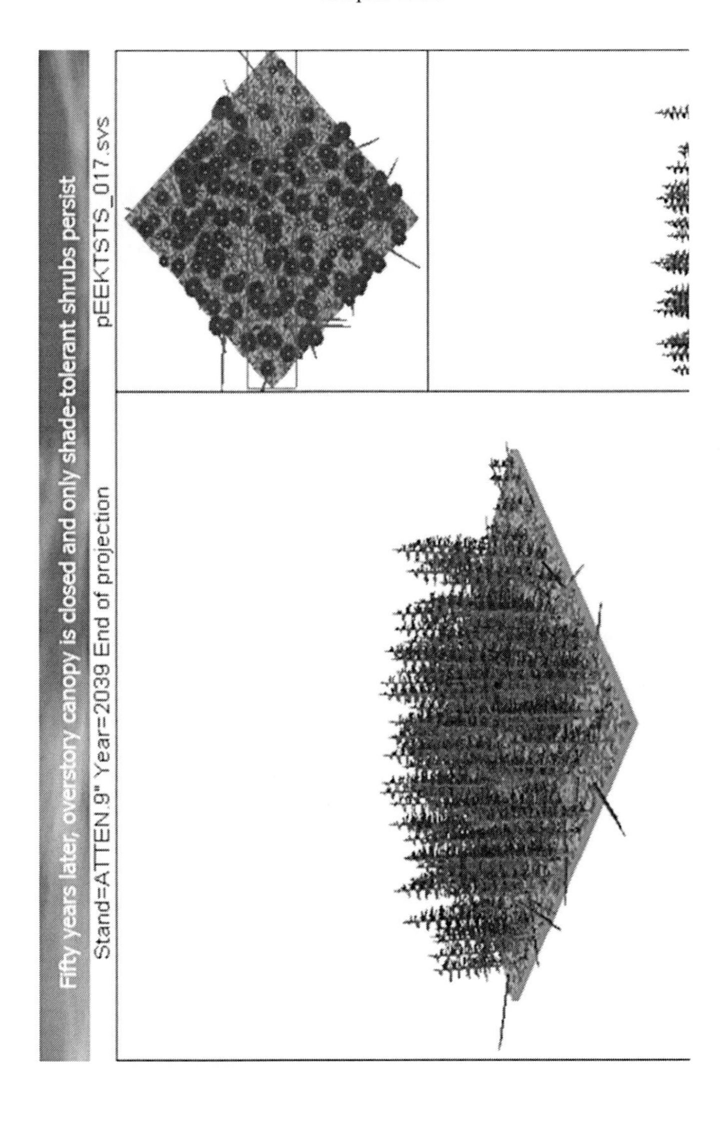
Fifty years later, overstory canopy is closed and only shade-tolerant shrubs persist
Stand=ATTEN.9" Year=2039 End of projection
pEEKTSTS_017.svs

There were four "notoriously bad wildfire years" in the 1900s: 1910, 1919, 1929, and 1934 (Space 1964). In 1910, 136,412 acres (55,204 ha) burned, in 1919,142,849 acres (57,809 ha) burned, in 1929, 65,275 acres (26,416 ha) burned and in 1934, 111,195 acres (44,999 ha) burned in the Lochsa (Figure 9-2). The Lochsa drainage comprises 789,355 acres (323,083 ha), (Norberg and Trout 1957 estimated 835,216 acres (338,000 ha). Over 582,665 (235,796 ha) had burned and/or reburned by wildfire since 1900 (Table 9-1). Norberg and Trout (1957) estimated 92,417 acres (37,400 ha) of winter range, of which 57,081 acres (23,100 ha) or 62% had been burned at least once. From Lowell to Lake Creek, 86% was burned, while above Lake Creek, 26% was estimated to have been burned. During their surveys, the numbers of elk on burned winter range was exceedingly high. The winter habitat use patterns that McLean (1972) described indicated that 98% of elk observed in December 1971, 75% in February 1971, and 79% in March 1971, were in burned-over areas. Use of prescribed burns and subsequent regrowth in the Pete King drainage during the study constituted 85% of the observations during that winter by 3 radiocollared elk using the drainage. Shrub field vegetation is without question a major winter habitat for Lochsa elk.

Space (1964) describes the spectacular start of the McLendon Butte fire of 1934:

> "This fire started in August about ½ mile northeast of McLendon Butte in a snag patch caused by a fire of 1917. It was a lightning fire reported at about 4:00 A.M. A CCC crew went to it but could not bring it under control. About 10:00 A.M. a snag burned off and fell and slid down the mountain towards Fish Creek. This scattered fire along a narrow strip for about 1/4 mile. Shortly thereafter this whole strip blazed up into a raging inferno. The fire swept back up the mountain to the southwest, widening and gaining force as it went. It reached the top of the mountain with such explosive force that it threw a spot fire across the Lochsa River 3 1/2 miles away. Another landed across Fish Creek near its mouth 4 miles away. There were hundreds of spot fires closer to the main fire. In a matter of an hour this fire had scattered over several thousand acres and set the stage for a major burn."

Table 9-1. Area burned by wildfire in the Lochsa before 1930, 1930-1950, 1951-1970,1970-1992.

Figures for Powell Ranger District are included for 2000-2012.

	HECTARES	ACRES	% OF TOTAL (<1930-1992)
AREA BURNED <1930	183094.9	452527.7	56.7
AREA BURNED 1931-1950	46910.4	115915.7	14.5
AREA BURNED 1951-1970	1479.9	3657.0	0.5
AREA BURNED 1970-1992	6045.4	14938.2	1.9
TOTAL	323091.1	798358.1	100.0
POWELL DISTRICT 2000-2012	68221.5	168,579	

Fire History

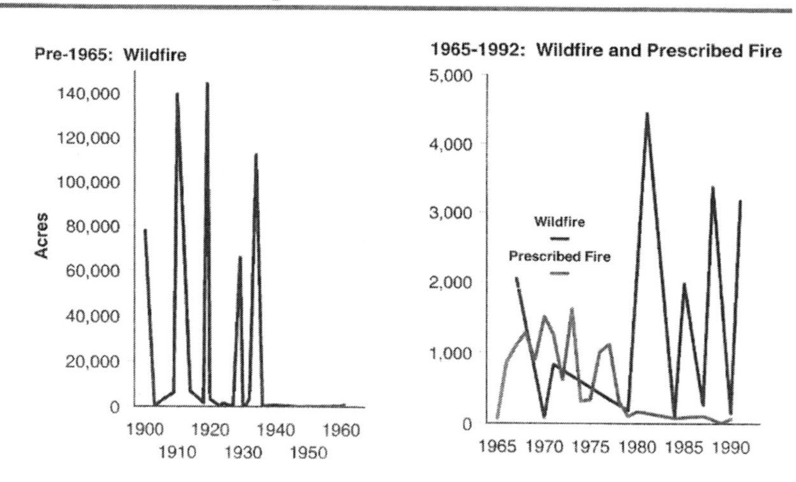

Figure 9-2. Fire history in the Lochsa including wildfire before 1965 and both wildfire and prescribed fire from 1965-1992.

Elers Koch was designated by the regional forester to manage fighting the 1929 Lochsa fire. His description of that fire, which started in August 1, 1929 from lightning strikes in and around Bald Mountain Creek, informs us that it was finally contained on the 8th of September after 480 men built approximately 95 miles of fire line (Koch 1998). After having a major role in fighting fires in western Montana and northern Idaho over the 1910-1934 period, Koch proposed that fighting fire within the Selway wilderness be abandoned (Koch 1935). The editor of the Journal of Forestry described Koch as "evidently a wilderness area enthusiast" alluding to this recommendation as being highly controversial. Regardless, 1,875,000 acres of the Selway-Bitterroot country was designated as a Primitive Area in 1936, which subsequently became the Selway-Bitterroot Wilderness in 1964.

However, Koch's views were not completely forgotten within the Forest Service. Moore (1996) reported that a recommendation to initiate controlled burns to provide forage for elk came about in 1948. However, it was not until 1965 when prescription fires were initiated in the drainage. Fire prevention had virtually saved the U.S. Forest Service from demise until the 1910 fires established the need for protection and the "Smokey Bear" campaign became the mantra for the agency (Egan 2009, Pyne 2008).

Fire suppression in the Lochsa drainage which was not incorporated into the Primitive Area became effective in 1940, a year of high fire occurrence but there was also a large number of trained firefighters. A total of 319 fires started that year, but only 765 acres (309 ha) burned (Space 1964). However, the exceedingly dry conditions which preceded the four large fires had not returned until 1981, when 4391 acres (1,777 ha) burned (Table 9-2).

The map showing wildfire distribution (Figure 9-3) portrays how the burns have occurred in the Lochsa. Fires before 1930, primarily the 1910, 1919, and 1929 burns, were widespread across the drainage except for its upper reaches. The Pete King and McLendon Butte fires of 1934 burned areas that had been previously burned, in the lower third of the drainage. Fires after 1950 have been largely inconsequential and mostly confined to the portions of the drainage located in the Selway-Bitterroot Wilderness.

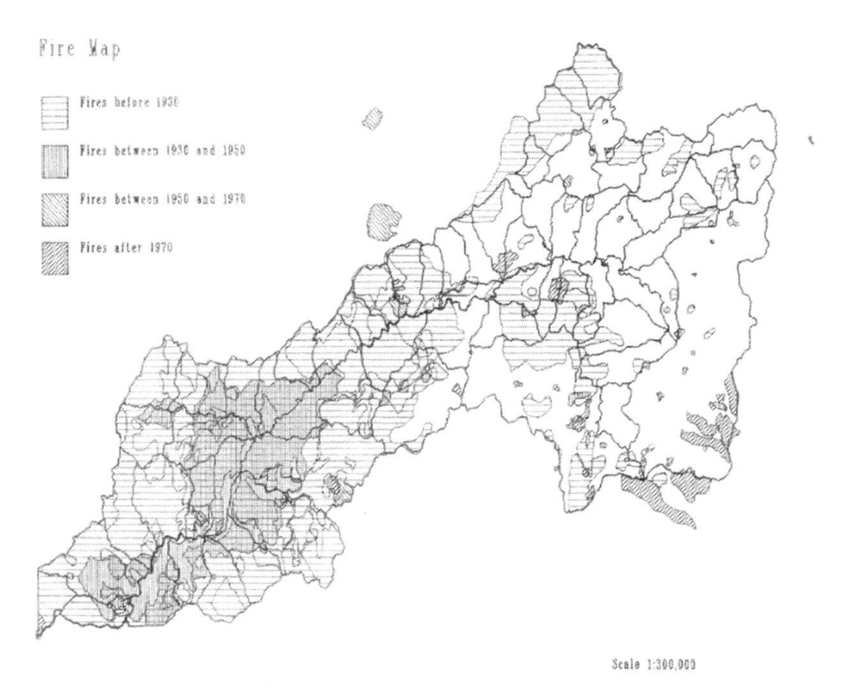

Figure 9-3. Distribution of wildfire through 1992.

Figures 9-4, 9-5, and 9- 6 show photographs of the Lochsa River drainage in 1936, just after the Pete King Fire of 1934. Figure 9-4, a U.S. Forest Service photo, shows north facing slopes were starting to regenerate to conifers, primarily Douglas fir, in the Fish Creek drainage, in 1980, 46 years after the 1934 fire. This is a major winter-spring range for elk and undoubtedly the browsing on southerly-facing slopes that are available in winter was limiting conifer regeneration. Figures 9-5 and 9-6 show the oblique aerial photos taken by the Washington Air National Guard in 1936. The inverse u-shaped bend in the Lochsa River identifies the locations. Small patches of conifers in the riparian zones were left at the head of Noseeum Creek, and a larger stand was unburned in Indian Grave Creek. The photos don't portray the regenerating shrub complex but do show how much of the areas were devoid of conifers following the fires. Figure 9-7 shows the extent of conifer regeneration in the Grave Peak area in 2016. Slopes without conifers will have tall shrubs, primarily Scouler willow and alder, that are mostly out of reach of browsing. Other slopes are on rocky outcrops and shallow soils that may not succeed to conifers or shrubs.

Figure 9-4. Fish Creek drainage looking southeast, 1980. Conifers are primarily on north-facing slopes. US Forest Service photograph., courtesy of University of Idaho Library.

Figure 9-5. Noseeum Creek 1936. Slopes are essentially devoid of conifers. U.S. Forest Service photograph taken by Washington Air National Guard, courtesy of University of Idaho Library.

Figure 9-6. Lost Creek (left [west] side of Grave Butte) and Indian Grave Creek on the right (east side), 1936. A large stand of conifer at the head of Lost Creek was left after the fire. U.S. Forest Service photograph taken by the Washington Air National Guard, courtesy of, University of Idaho Library.

Figure 9-7. Lost Creek, Grave Peak, and Indian Grave Creek, 2016. Darker vegetation is sapling and mature timber. Peek photograph.

Estimates of vegetative categories representing different stages of succession were developed for 1972, 1991, and 2012 (Table 9-2, Figure 9-8). The information suggests that after conifers, primarily Douglas fir, overtop the shrubs, conifer development proceeds much faster. This is shown in the decline of the grass-forb-shrub component to 3.6% of the 1972 acreages by 2012. This component was reduced to 62 % of the1991 estimate by 2012. The shrub-sapling component was 20.7 times the estimate in 1972 by 2012. Even-aged timber and mature timber had been reduced, primarily by the logging in the upper parts of the drainage, but also by wildfire. Nonvegetated area had declined between 1972 and 1991, but then did not change much from 1991-2012 and the changes in this type may be attributable to the differences in methodology as much as what was occurring on the ground. The number of stands described as shrub-sapling, even aged timber and mature timber increased comcomitant with the estimates for the area changes as well. The estimates don't reveal species composition in the herbaceous and shrub/sapling stages, which will reflect the browsing pressure over the years by the elk. The information shows how dramatic the vegetation changes have been in the drainage.

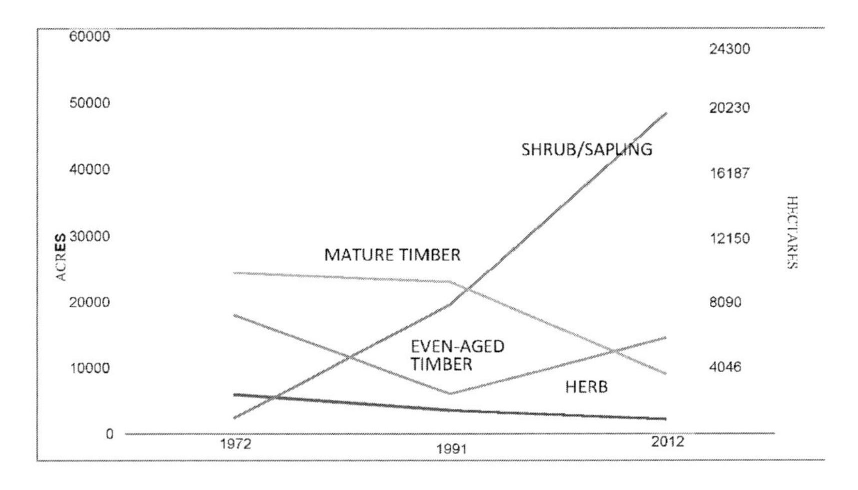

Figure 9-8. Area in four different vegetation types in the Lochsa drainage. The shrub-sapling stage consists of tall shrubs interspersed between conifer saplings, all usually over 4.5 meters (15 feet) tall and of little or no forage value for elk. Mature timber decline reflects the harvests in the eastern portion of the drainage while even-aged timber is closed-canopy forest with little understory growth. Even-aged timber represents regenerating forest that lacks the shrub component in shrub-sapling forest.

Table 9-2. Estimates of vegetation composition in the Lochsa River drainage in 1972,1991, and 2012. These data are taken from GIS procedures described earlier and were provided by Chang for 1972 and 1991, and Aycrigg for 2012.

Habitat	Number of stands			Area acres (hectares)---------		
	1972	1991	2012	1972	1991	2012
Non-vegetated	1048	834	917	28,589 (11,570)	17,146 (6,939)	18,484 (7,483)
Grass-forb-shrub	1666	1071	1278	610,511 (247,070)	35,424 (14,336)	21,865 (8,852)
Shrub-sapling	693	1668	5152	23,403 (9,471)	195,394 (79,075)	484,660 (196,219)
Even-aged timber	2410	2053	5441	180,526 (73,058)	60,626 (24,535)	144,953 (58,685)
Mature timber	1530	1649	3653	245,555 (99,375)	230,771 (93,392)	90,685 (36,715)

CHAPTER TEN

SHRUB ECOLOGY INVESTIGATIONS

As might be expected from their importance, the ecology of the seral shrub fields which developed following fires has been extensively investigated. In the cedar-hemlock and grand fir forests typical of northern Idaho. The tall shrub complex that develops after a site is burned may have many species which were present only as seed or scarce, decadent, and tall individuals in the mature conifer forest. The shrub complex is comprised of many shade-intolerant shrubs as well as shrubs that occur under tree canopies. All of these shrub species will resprout from underground roots or rhizomes although some resprout more vigorously than others.

The herbaceous understories in these shrub fields are comprised of forbs and grasses which are also capable of surviving in open sunlight with competition from the woody species. Maximum woody forage production of the shrub complex varies from 6 to 30 years after wildfire, which in the Lochsa means that the shrub fields in the middle and lower areas last burned in 1934 probably reached maximum production in the early 1940s through early 1960s. As shrub fields are gradually invaded by conifers, conifers begin to shade out the shrub community, and the sapling-pole stage develops which has fewer shrubs and mostly shade-intolerant herbs.

Mueggler (1965) documented the effects of environmental variables on the growth and composition of these shrub communities. His studies were north of the Lochsa drainage in the North Fork of the Clearwater, St. Joe, and Coeur d'Alene River drainages, but provide an indication of the situation in the Lochsa. The habitat preference of nine shrubs commonly browsed by elk and other big game are summarized in Table 10-1.

Perhaps the most significant relationship from the standpoint of the effects of fire is the highly variable ability of different shrub species to persist under tree canopies. The two ceanothus species, snowbrush ceanothus, and redstem, are very shade intolerant and don't survive for long as conifers overtop them. Redstem is the most highly palatable shrub in the region. Mueggler (1965) did not find snowbrush ceanothus in unburned areas, and redstem was scarce. Bitter cherry also grows predominantly in areas where the conifer overstory is absent. On the other hand, mountain maple, Scouler willow, and mountain ash persist in regenerating conifer stands and in forest gaps in older forest. Serviceberry and Utah honeysuckle do not appear to be appreciably affected by changes in overstory cover. Pachistima (boxwood) shows a preference for partial overstory cover, but will respond also to fire by increasing its cover from that which occurs in closed-canopy mature forest.

Thus, the species composition of shrubs will change as the sites become invaded by conifers which begin to overtop the lower shrubs and then compete for light and nutrients with taller species as well. Species like bitter cherry and the two ceanothus species will decline as they become shaded, while pachistima may increase as the site becomes partially shaded, and finally the stand will consist of honeysuckle, serviceberry, Scouler willow, mountain maple and other species which can persist with the conifers. If the stands are heavily browsed, unpalatable species like ninebark, oceanspray, and snowberry may dominate in the stand as the more palatable species are weakened from the browsing. White birch and alder will also persist since they can grow out of reach of browsing and can persist with partial shade. Since Douglas fir is a palatable species, sites on which this conifer is regenerating may persist in the shrub-dominated state with the less palatable species prevailing, for longer than if heavy browsing pressure is absent.

Shrubs which persist in sapling-pole and older stages are often tall, spindly and unproductive of forage for elk. Mountain maple, serviceberry and bitter cherry grow beyond 2.1m (7 feet) tall, the height generally considered to be out of reach by elk, in 31-40 years (Mueggler 1965). Scouler willow, a fast-growing shrub which is palatable, can grow to that height in a decade or less. The lower-growing shrubs remain essentially in reach of elk throughout their life in the shrub field.

Table 10-1. Habitat preferences of selected tall shrubs that are palatable to elk (Mueggler 1965).

Species	Soil	Tree Canopy	Grows out-of Reach in	Slope	Exposure
Bitter cherry	High K.	Open canopy	31-40 yrs	mid-slope	south
Mountain Ash	High organic matter	Persists in closed canopy	No	No pref.	No pref.
Mountain Maple	High K	persists in shade	In 31-40 years	No pref.	N & S
Pachistima	High organic matter	Partial overstory	No	No pref.	No pref.
Redstem	High K Low organic matter	no canopy preferred	No	mid-slopes	south
Serviceberry	No Pref.	No Pref.	In 31-40years	Lower slopes	south
Snowbrush	moderately deep soils 14-20”	no canopy preferred	No	No pref.	south
Scouler willow	No pref.	Highest in open, persists	21-30 yrs	Lower slopes	No pref.
Utah honeysuckle	No pref.	reduced on burned areas	No	pref.	No pref.

There are other habitat relationships which influence the composition and density of a shrub field. Mountain maple, redstem, and bitter cherry grow best on soils with high potassium levels (over 34 g/m^2). Pachistima and mountain ash grow best in soils of high organic matter content. Snowbrush ceanothus grows best in moderately deep soils of 36-51 cm. Snowberry and Scouler willow will be found more often on lower slopes, while redstem appears to express itself best at mid-slope. Serviceberry, redstem, snowbrush ceanothus, and bitter cherry grow best on southerly exposed slopes. Finally, the huckleberries and pachistima show increasing amounts of cover at higher latitudes in Idaho, while snowberry is more abundant further south, which means pachistima is less important in the Lochsa than further north.

Mueggler (1965) considered shrub fields notoriously difficult to classify because they exhibit high variability in species composition under similar environmental conditions. However, there were correlations between species associations that were apparent. Serviceberry and mountain maple, and snowbrush and bitter cherry were apt to be associated with each other. Redstem was associated with ninebark, bitter cherry, Scouler willow and huckleberries. Mueggler (1965) concluded this comprehensive investigation by stating that pachistima, huckleberries, honeysuckle, thimbleberry and mountain maple were the most common shrubs under relatively closed tree canopies. Under open canopies, Scouler willow, and the species common under closed tree canopies predominated. Scouler willow, bitter cherry, redstem and snowbrush decrease in importance as the tree canopy closes. Major changes in herbaceous species composition also occur relative to canopy cover.

In the Lochsa, redstem ceanothus was considered the most palatable forage-producing shrub. This species is common in brush fields at middle to low elevations where elk winter and comprised 68% of the palatable shrubs in a survey of five Lochsa River locations in 1950 (Weyerman et al. 1950), and 49% of the palatable shrubs in a survey of Sherman Creek in 1960 (Williams 1962). Redstem does not produce many sprouts unless the above-ground portion of the plant is severely damaged by fire or cutting (Hickey and Leege 1971). In an experiment involving ten plants that were burned in Fish Creek in late March which averaged 3.4 sprouts per plant before the fire, an average of over 35.4 sprouts occurred after the fire. Subsequent investigations suggested that slightly more sprouts were produced after spring burning than fall burning. However, fall burns resulted in establishment of more seedlings than did spring burns, which is a consideration if an area needs to be restocked with this species (Leege 1969).

Redstem seeds have a small black impermeable coat that germinates when treatment causes the coat to become permeable to water. Fire or abrasives of one sort or another may cause germination, and fire is of course the most common means of germinating in nature. Ceanothus seed can persist for extended periods of years in the duff and soils under mature forest canopies until a fire removes the canopy and causes the seed to germinate (Gratkowski 1962, Quick and Quick 1961, Hickey and Leege 1970).

Redstem also withstands browsing well in spring and fall as well as in winter (Young and Payne 1948). A six-year investigation showed that spring or fall removal of 75% of the annual growth of redstem could be sustained without significantly affecting production. However, continuous and heavy browsing of new growth during summer can severely retard and even eliminate newly-established redstem plants (Leege 1984). Redstem mortality due to browsing also may be important in reducing the numbers of sprouts the first growing season following burning (Hickey and Leege 1970). These researchers reported that redstem generally remains available to elk, remaining below 2.1 m in height. The review of redstem ecology by Hickey and Leege (1970), and subsequent extended research into productivity of this and associated shrub species, attests to the importance attached to the species and to shrub fields for elk.

Redstem and snowbrush ceanothus are susceptible to dieback due to late frosts and sudden drops in temperature when exposed above the snow (Young and Payne 1948, Stickney 1965). Redstem production was related to May-August precipitation, with annual twig growth averaging 24-30 cm depending on moisture. Old plants growing on east-facing slopes appeared to be most susceptible to dieback, and researchers suspected that many shrubs and flowering plants had already been eliminated on north slopes. Leege and Hickey (1975) concluded that dieback caused by extreme temperature fluctuation acts as the mechanism for removing redstem from a plant community when plant vigor has been reduced by shading from conifers and taller shrubs.

Shrub Productivity

Information on productivity of the shrub fields at the height of their development following the earlier wildfires was not obtained. Earlier surveys were primarily directed at estimating utilization, while Mueggler (1965) used estimates of cover to arrive at species composition and trends. Biomass estimations are expensive of time and effort, both of which are always scarce. However, Williams (1962) explored use of distance measures to arrive at plant density estimates as a means of quantifying changes in the shrub fields. Estimates for a southwest exposure in a Pete King Creek shrub field were 1.8 shrubs/m^2 (7284/acre), of which redstem comprised 21.8% of that composition, or 0.44 plants/m^2 (1780/acre). Subsequently, Leege et al. (1972:69) reported that annual production of unbrowsed plants averaged 304 grams (2.87 lbs) per plant over a four-year period. Alldredge et al (2001) reported production of 49.9gm/m^2 (445.1lbs/acre) on broadcast burns on north-facing slopes 10-20 years after logging (Table 10-2).

If we assume that these figures are representative of redstem growing in shrub fields at their period of highest production, then redstem alone could have produced 119 g/m^2 (1061 pounds/acre) of forage after leaf fall. A Scouler willow plant 3 m tall and 2.4 m in diameter was estimated to produce 765 grams (1.68 lbs) of growth (Landall and Fager 1935). If willow comprised 10% of the composition or 0.22 plants/m^2 (890/acre), then over 165 g/m^2 (1392 lbs/acre) of forage of willow could have been available. While these estimates are quite likely high, it is apparent that these shrub fields may have been producing between 112-225 g/m^2 (999-2007 lbs/acre) of browse at their maximum period of productivity when all forage species are considered and snows didn't preclude availability. These estimates involve tenuous assumptions, but the estimates may be compared with estimates of 34-56 g/m^2 of forage on bunchgrass winter ranges and to estimates of 10-38 g/m^2 of forage on a site burned in spring between 6 and 11 years later (Tables 10-3, 10-4).

The Selway data set provided insights into shrub composition and density that showed how important recently burned areas were for producing species that were palatable to elk (Peek et al.1992). There was no appreciable presence of redstem ceanothus or Scouler willow on the older burns or the unburned stands. Highest seedling densities of both species were in the 1-6-year old stands with redstem being highest on north-facing slopes at the lowest elevations and Scouler willow highest on north-facing slopes at the highest elevations.

Soil nitrogen levels doubled at 5 years following the fires, coinciding with highest levels of redstem and pellet distributions (Figure 10-1). The implication of this is that redstem, a nitrogen fixer, contributed to increases in soil nitrogen which may facilitate subsequent establishment of conifers on these sites.

The 1-2-year-old burns had some Douglas fir and white birch seedlings at the 3000-3500 foot (900-1500 m) level. Syringa, ninebark, and ocean spray, all unpalatable shrubs, were also present. Mountain maple and serviceberry, are taller growing shrubs that are palatable and were also present. The 5-6-year old burns contained grand fir, syringa, ninebark and mock-orange, which are not palatable. Besides Scouler willow and redstem, mountain maple, serviceberry, snowbrush ceanothus, and pachistima were present, all palatable. Conifers dominated the 50-100-year-old stands with Douglas fir being most abundant with grand fir, western red cedar, white birch and Ponderosa pine also present. Mountain maple, a tall-growing species that can reach into the canopies of the conifers was the most common shrub, with some of the other species mentioned above being present.

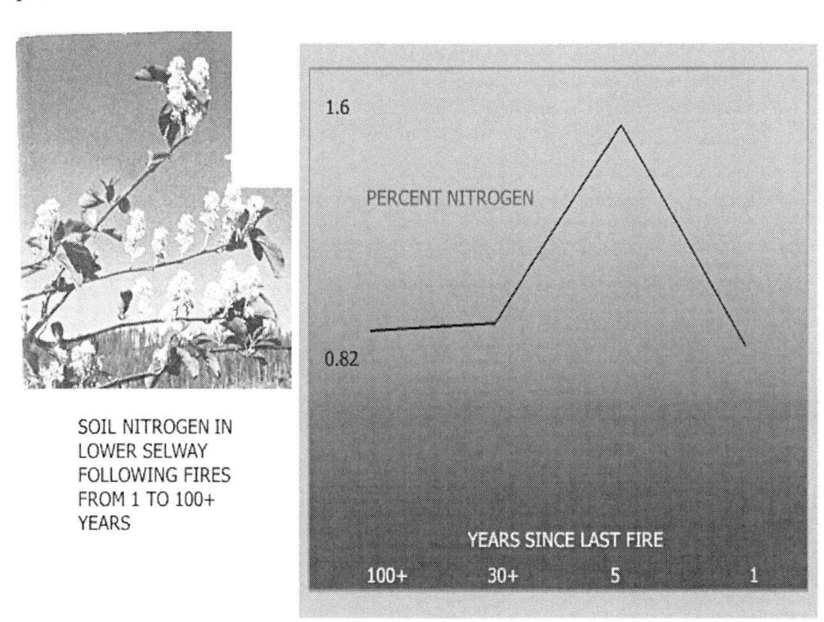

Figure 10-1. Soil nitrogen doubled at 5 years post-fire on the lower Selway study sites when compared to older and younger stands. These were fall prescribed burns. Redstem ceanothus in flowering stage benefitted. Modified from Green and Talbert (1991).

Syringa, ocean-spray, and serviceberry, species that tolerate more shade than Scouler willow and redstem, were consistently present in the stands that were examined. The unburned sites had high amounts of grand fir with Douglas fir, western red cedar, and Ponderosa pine being present. Mountain maple and syringa were present in the stands. The change from palatable shrubs being present in the recent burns to the less palatable shrubs dominating in the older burns and unburned forest was obvious. While this change can be primarily attributable to the increase in conifer cover that over shadowed the understories, elk browsing also undoubtedly enhanced the change.

Western red cedar/pachistima habitats are common on the lower elevation forests west of the Bitterroot Mountains that includes the Lochsa River drainage. These forests are highly productive and have been logged since the early 1900s, resulting in a mixture of stand types and age classes. Alldredge et al. (2001) sampled 77 stands representing the primary types in 1997 and 1998 on highly productive soils west of the Lochsa River drainage. Sampling of stands aged 1 to 20 years showed large quantities of forage suitable for elk, much as the lower Selway and the Lochsa studies showed. This work provided estimates of rooted stem densities and current year's growth of twigs for shrubs in the different stand age classes (Table 10-2). Redstem ceanothus was the most abundant shrub in stands that were 20 years old or less. Pachistima was also important in naturally regenerating sites. Pachistima is more shade-tolerant than redstem and persisted in the older sites that were dominated by conifers. Both species are highly palatable for elk. Other species that are palatable to elk occurred in smaller amounts, with mountain maple and serviceberry persisting in mature timber. Scouler willow persisted on north slopes into small timber stands, but not into the larger stands. Syringa, alder, and ninebark persisted into mid-successional stands but not into the small and large timber stands. These three species are not very palatable to elk.

Annual biomass of shrubs on south aspects reached highs of 256.93 pounds per acre (28.8 gm/m^2) and then dropped off to less than 20 % of that estimate in the timber categories. On north aspects, biomass reached 445.2 pounds per acre (49.9 gm/m^2). These figures compare with biomass estimates for bluebunch wheatgrass on south exposures in the Middle Fork Salmon River that average 461.3 pounds per acre (Peek 2014). However, the shrub estimates include both palatable and unpalatable species while bluebunch wheatgrass is a highly palatable elk forage.

Table 10-2. Densities and biomass of selected shrubs in different ages of forests immediately west of the Dworshak Reservoir in north-central Idaho. (Alldredge et al. 2001).

Forest age class	1-20 years	Mid-succession	Small timber	Large timber	1-20 years	Mid-succession	Small timber	Large timber
	-----------	SOUTH	ASPECT	------------	------------	NORTH	ASPECT	-----------
MOUNTAIN MAPLE	0.07/2.59[1]	0.05/0.48	0.14/0.83	0.19/0.7	0.09/2.59	0.08/4.36	0.05/0.29	0.07/2.8
THINLEAF ALDER	0.01/8.28	---	---	---	0.21/8.28	0.01/0.05	---	---
SERVICEBERRY	0.05/2.45	0.10/4.65	0.16/6.24	0.09/1.8	0.06/2.45	0.15/7.01	0.01/0.14	0.13/0.24
REDSTEM CEANOTHUS	0.13/4.84	0.06/1.91	---	---	0.07/2.01	0.05/0.78	0.01/0.14	0.13/0.24
SHINEYLEAF CEANOTHUS	0.08/---	---	---	---	---	---	---	---
OCEAN SPRAY	0.05/4.4	0.01/1.03	---	0.01/0.2	0.05/1.28	---	---	---
PACHISTIMA	1.05/47.9	0.36/7.75	0.65/3.84	.28/0.67	0.42/7.26	0.18/2.47	0.39/1.63	0.14/0.91
NINEBARK	0.17/5.83	0.04/5.58	---	---	0.40/6.89	0.01/0.94	---	---
BITTER CHERRY	0.02/0.91	0.01/0.07	---	---	0.02/0.19	---	0.04/0.14	---
ELDERBERRY	0.07/0.21	0.04/0.36	---	---	0.07/0.80	0.02/0.04	---	---
SCOULER WILLOW	0.02/1.13	0.02/3.36	----	----	0.03/3.42	0.02/0.49	0.01/0.11	---
MOUNTAIN ASH	0.01/0.06	0.01/0.05	---	---	0.02/0.19	0.01/0.30	---	---

[1] STEMS/M^2/TWIGS/M^2. Stems are rooted; twigs represent current year's growth.

	1-20 years	Mid-succession	Small timber	Large timber	1-20 years	Mid-succession	Small timber	Large timber
	-----------	SOUTH	ASPECTS	------------	------------	NORTH	ASPECTS	------------
ANNUAL BIOMASS	BROADCAST BURN							
STEMS (gm/m²/lb/acre)	24.3/216.8	28.8/256.9	4.1/36.6	5.2/46.39	49.9/445.2	21.1/188.25	4.5/40.15	2.7/24.09
TWIGS(lbs/acre/ kg/ha)	216.8/243.5	256.9/288.5	36.6/41.1	46.4/52.1	445.1/498.9	188.2/211.3	40.1/45.0	24.1/27.1

Further evaluations of forage quality were conducted by Alldredge et al. (2002) west of the Lochsa on industrial forest lands. These examinations extended from May through November, when elk were using these areas. Species included clover, western goldthread, elk sedge, Kentucky bluegrass, redstem ceanothus, Scouler willow and serviceberry, all present in the Lochsa drainage as well. Crude protein and digestible energy levels declined progressively from May through November with the three shrub species generally remaining higher than the forbs and grasses. The values of all forage species were below requirements from June through November (Cook et al. 2004), suggesting that mixes of forage species are necessary, and selection of the most nutritious plant parts prevails.

Production of current-years-growth forage on spring prescribed burns at Holly Creek from April 1966 when the site was burned until October 1977, twelve growing seasons later (Table 10-3). While production varied considerably between years, it was still higher at year 12 than before the burns. Willow had again grown out-of-reach, but redstem especially continued to be more productive. Serviceberry did not respond appreciably to burning as would be expected from Mueggler's (1965) investigations. Heavy utilization of redstem was noted in summer and that decreased the vigor and browse production of some plants.

The investigations of spring burning on Polar Ridge in the Pete King drainage provided some idea on response to burning different aspects (Table 10-4). Generally, production on south slopes was much greater over the four year period after spring burning, with redstem again being the species that responded best. Scouler willow responded best on a west-facing slope but grew beyond reach rapidly.

Table 10-3. Available production of willow, redstem, serviceberry and mountain maple on a site burned in 1966 on Holly Creek (Leege 1978a).

	Current year's growth in lbs/acre (gm/m^2)				
SPECIES	Scouler Willow	Redstem	Service berry	Mountain maple	Total
Prior to burning 1966	43.5(4.87)	20.4(2.29)	14.8(1.65)	0.3(0.03)	79(8.85)
Year after 1967	171.5(19.22)	41.6(4.66)	13.8(1.55)	0.6(0.07)	222.5(24.94)
Six growing seasons after (1971)	45.5(5.10)	36.9(4.13)	6.6(0.74)	0.5(0.06)	89.5(10.03)
Seven growing seasons after (1972)	40.5(4.54)	87.5(9.84)	6.2(0.69)	0.8(0.09)	135.0(15.13)
Nine growing seasons after (1975)	14.6(1.64)	84.2(9.44)	1.7(0.19)	0.6(0.11)	101.1(11.33)

Table 10-4. Available production of willow, serviceberry and mountain maple on Pete King drainage (Leege 1978b). Current years growth is in lbs/acre (gm/m²).

	Scouler willow		Serviceberry		Mountain Maple[2]	Total	
	West	South	West	South	South	West	South
1967[1]	6.7(0.8)	9.1(1.0)	1.1(0.1)	7.0(0.8)	4.1(0.5)	14.8(1.7)	88.5(9.9)
1969	296.3(33.2)	65.4(7.3)	0	3.7(0.4)	7.9(0.9)	298.1(33.4)	131.7(14.8)
1970	120.1(13.5)	14.4(1.6)	12.0(1.4)	8.2(0.9)	11.8(1.3)	137.2(15.4)	159.0(17.8)
1971	74.9(8.4)	10.3(1.2)	6.9(0.8)	5.0(0.6)	20.3(2.3)	109.9(12.3)	338.0(37.9)
1972	11.5(1.3)	1.0(0.1)	2.5(0.3)	6.0(0.7)	5.2(0.6)	43.3(4.9)	265.1(29.7)

[1] Data taken before site was burned.
[2] Not present on west slopes

Planned ignitions of winter ranges occurred continuously from 1965 through 1980. The area burned peaked in 1973 at 637 ha but burning became intermittent after 1980 with burning occurring only in five years since (Figure 10-2). Burning principally occurred on lower elevation winter ranges along the Lochsa River corridor. The acres burned were obviously insignificant compared to acres burned by wildfires. Annual average for the 26 years, 1965-90 was 189 $\pm$ 212 ha. This compared to average annual 2,002 $\pm$ 8,088 ha burned by wild fire during the 92 years since 1900. Or compared to the 1965-90 period which for wild fires burned 211 $\pm$ 457 ha annually.

Figure 10-2. A spring burn in an old shrub field, illustrating the low intensity of the fire. Idaho Department of Fish & Game photo.

However, during the height of the prescribed fire program for improving elk winter range between 1966 and 1977, an average of 997 acres (400 ha) was burned annually. The burns were primarily spring fires that were easily controlled, but for which the flammability of the shrub fields between snow-melt and initiation of growth was very short, estimated to average 4 days each year. Seasons which did not produce much acreage were years of wet springs when the "burning window" was short or absent. During the 12-year period, five years produced areas below the average, including two years (1974, 1975) of only 126 and 134 ha burned. Efforts to increase the areas burned during the few good days resulted in the use of aerial ignition techniques (Leege and Fultz 1972).

IDFG and the Clearwater National Forest cooperated on this prescribed burning program with the IDFG conducting research on fire effects and the USFS providing manpower and equipment. However, in the late 1960's it was believed that additional funding was needed to get the appropriate area burned. Beginning in 1970 and for several years, the IDFG contributed large sums of money to the program until it became obvious that other factors were restricting the growth of the program. Worries over burning regenerating conifers and about burning mature timber stands above winter range inhibited the burning program.

In later years, prescribed fire activity dropped, averaging only 30 ha between 1978 and 1991. Wet springs, doubts about the value of the program, complaints about burning up young conifers, and complaints about smoke pollution by the public all entered into the reduction in acreage burned.

Earlier attempts to use herbicides to kill aerial portions of out-of-reach shrubs and promote resprouting proved unsatisfactory, primarily because redstem ceanothus was completely killed by the treatments (Mueggler 1966, Lyon and Mueggler 1968). Leege (1969) reported that the first use of prescribed fire was in the St. Joe National Forest in 1961 and had produced desirable results.

Several aspects of shrub field responses to prescribed fire were investigated. Comparison of spring and fall burns were made, nutrient content of regrowth and unburned twigs of key shrub species was compared, as was productivity of newly burned shrub fields with unburned ones. Aerial ignition techniques were evaluated as efforts to decrease costs and increase areas burned were made. Concerns over erosion, invasion of bracken fern, effects on understory composition, and effects of repeated burning of the same site were investigated. The use of plant regrowth by elk after burning was also evaluated as was the technique of thinning young conifer stands to retain browse (Leege 1984). These investigations extended over a 13-year period and provided substantial information about the use and results of prescribed burning in the region. Since the information appeared in widely read publications, the project received extensive interest across the country and was a model for large-scale wildlife habitat improvements using fire in mountainous terrain (Leege 1968, 1969, 1971).

Two experimental burns were conducted in 1965, on Otterslide Creek and Sherman Creek, to evaluate sprouting responses of 11 shrub species to spring and fall burning (Leege and Hickey 1971). Only 261 of 366 marked plants had aerial portions killed by fire, illustrating the typical pattern of burning wherein treatment areas are incompletely burned. Redstem ceanothus, bitter cherry and cascara showed some minimal mortality, particularly from the fall burning. Of 126 marked plants burned in the fall, 7 (5.6%) failed to sprout. Only 2 of 135 spring burned shrubs (one redstem and one bitter cherry did not sprout. There was a 3.7% mortality of bitter cherry and redstem on control plots, illustrating that fire was not the only cause of mortality. Leege and Hickey (1971) thought that shading and competition with other plants were contributors to mortality besides fire. Plants which were burned in late March and early April started to sprout within 4 to 8 weeks and provided browse the winter after treatment, an advantage for burning in spring. While there were more sprouts produced from spring burns, sprouts from the fall burns tended to be longer. Crown volume (Volume $=\prod r^2 h$ where r is the radius of the preburn crown and h is the height of the plant) was significantly correlated with numbers of basal sprouts produced from burning.

While production increases were demonstrated with resprouting shrubs, most particularly redstem, whether seedlings would establish and grow also was a source of interest. Redstem was again of particular interest because it had declined with advancing age of the shrub fields. Investigations by Hickey and Leege (1971) indicated that fall burning produced more seedlings than spring burning, which would be expected if seeds were adapted to germinate following hot fire typical of the prevailing fire regime in the region. Late summer and fall fires were the rule, while burning during a very short window when plants were flammable in spring was rare if not absent in nature. The investigations suggested that as many as 96,800 redstem seedlings were produced in fall as compared to 24,000 in spring on burns, on a per hectare basis. Both seasons obviously produced large numbers of seedlings. The Holly Creek investigation suggested that many seedlings could survive, even in the face of browsing pressure by elk (Leege 1978a). However, germination was related to soil temperature during burning. Heat treatments of 982 C or more is preferred to germinate redstem, and subsurface temperatures on spring burns of less than 816 C occurred at 23 of 120 sites measured by Hooker (1972).

Orme and Leege (1976) summarized results from 12 study sites in the Lochsa. They found that redstem seedling survival and growth were better after fall burns than spring burns and that summer precipitation, aspect and elevation also influenced survival. Above normal precipitation, north and east aspects, and elevations of 900 m versus 600 m all produced better seedling survival. But not even fall burning guaranteed redstem establishment because exceptionally dry summers were successful in eliminating almost all of the newly emerged seedlings.

While production increases are a goal of burning shrub ranges, changes in nutrient levels of twigs in regrowth following burning also were investigated. Results of the studies in the Lochsa suggested remarkably little change in nutrient concentrations in burned and unburned twig ends (Table 10-5). A hot late summer burn in the upper Selway did increase nutrient concentrations, suggesting that spring burns may not encourage increased nutrient content in twigs the following year since soils would still be cool and plants would have a growing season after the fires in the same year.

However, the increase in sprouting on burned areas still makes more nutrients available on these sites for a period following burning, which may last for 10-20 years on sites burned with prescription fire or wildfire. These results are different than reports of increased nutrient concentrations following burning elsewhere and may be related to soil fertility, soil moisture, as influenced by severity of the burning.

Table 10-5. Concentrations of nitrogen, calcium, and phosphorus in twigs in unburned and burned sites in the Selway and Lochsa Rivers (Merrill et al. 1982, Leege 1969).

		REDSTEM			SERVICEBERRY		
		N	Ca	P	N	Ca	P
Lochsa							
	unburned	1.5	1.28	0.32	1.29	1.28	0.29
	spring after	1.6	1.29	0.28	1.63[2]	1.36	0.3
	fall after	1.5	1.23	0.19	1.58	1.26	0.23
Selway[1]	unburned	1.5[2]	1.2	0.15	1.2[2]	1.1[2]	0.37
	burned	2.2	1.4	0.16	1.4	1.2	0.31

[1] Burned sites are 3-year averages following burning.
[2] Differences between burned and unburned are signficant at $P<0.05$.

Concern over continuing to reburn shrub fields at regular intervals was raised early on, since "blowouts" of streams, including Fish Creek, following the major fires resulted from destruction of ground cover and extensive erosion. Undoubtedly some of this concern arose from those who would have preferred seeing these shrub fields restored to timber, but subsequent controversies over the effects of timber harvest and road construction on soil loss served to provide some perspective on the shrub field management program. Leege (1979b) investigated a shrub stand that was treated with three burns at five-year intervals. Redstem ceanothus, the "target species", increased in density through establishment of new seedlings after each fire. Leege observed redstem plants producing seed two years after burning, so a supply of seed on a site burned at five-year intervals was assured. However, sprout heights and numbers for other species, especially bitter cherry, tended to be reduced, and a gradual decline in density may occur for other species if burns are repeated at five-year intervals, especially when they are actively growing. Leege (1979b) felt that if burns were conducted at a 10-15-year interval, little change in shrub composition and density would occur. Increases in burn frequency may be used where redstem is present but scarce and a goal of increasing its density is indicated.

Leege and Godbolt (1985) found no evidence that soil erosion was increased following burning. On a site that had 7% bare ground prior to burning at the mouth of Noseeum Creek, on a south aspect with a 40-80% slope, bare ground increased to 10% the first year following fire, then declined to 4% the 2d year and 4th year after the fire. The increase of the second year was slight and lasted for only one season and was subsequently followed by declines in bare ground.

On the Sherman and Otterslide Creek study areas, significant increase in bare ground occurred following burning, and lasted for at least 5 years (Table 10-6). Bare ground increased from 5% to 40% after spring and fall burning, returning to preburn conditions 5 years later on the spring burned sites. Bare ground was still higher on the fall-burned sites at 7 years on one site but had returned to pre-burn levels at 6 years on the other site. No obvious changes in erosion were observed on either spring or fall burns. The decomposed granites of the Idaho batholith that characterize the substrates in the Lochsa are highly erosive if disturbed by roads (Bethlamy 1967), but these sites may not be altered sufficiently by prescribed burning to cause noticeable soil movement. Hooker's (1972) investigations of the prescribed fires in Holly Creek suggested that burns of low intensity that are typical have little effect on soils and plants (Figure 10-2). When soil temperatures occur over 538 C°, grasses and grasslikes are adversely affected and thimbleberry is benefitted. Soil nitrate nitrogen and nutrient content is benefitted from the typical cooler burns, which also are not of high enough intensity to cause losses of ammonium nitrate. Hooker (1972) observed conditions which suggested active soil movement from the earlier wildfires of 1910 and 1934, although the effects of those fires "can only be imagined".

Table 10-6. Changes in percent bare ground following prescribed fire in Otterslide and Sherman Creeks for the 1965-1972 period (Leege and Schlegel 1973).

PERIOD	FALL BURN	FALL BURN	SPRING BURN	SPRING BURN	CONTROL FOR SPRING BURN	
	SHERMAN	OTTERSLIDE	SHERMAN	OTTERSLIDE	SHERMAN	OTTERSLIDE
PREBURN	3	7	1	13	4	6
YEAR 1	42	37	20	53	-	8
YEAR 2	36	21	23	27	5	17
YEAR 3	19	30	13	19	3	8
YEAR 4	17	28	12	10	3	8
YEAR 5	0	18	5	8	-	6
YEAR 6	6	-	-	-	-	-
YEAR 7	-	20	-	-	-	-

Because winter range may also serve as spring range for elk, and herbaceous plants appear in the diet as they become available, Leege and Godbolt (1985) investigated responses of herbs to fire in the Noseeum Creek area. In this investigation, a control, a burn, and a burn and seed treatment were evaluated. The burn-seed treatment included broadcast seedings of slender wheatgrass, crested wheatgrass, orchard grass, bulbous bluegrass, tall wheatgrass, yellow sweet clover, and big bluegrass. Of the 40 nonseeded species observed, seven increased, eight decreased, and the remainder showed no trend (Table 10-7). Of the seeded species, only 4 were present in the plots after four growing seasons, with orchard grass, a species palatable to elk, being most abundant. Bracken fern production was favored by burning. The results suggested that if spring elk forage is to be improved, broadcast seeding with a mixture of 25% orchard grass, 25% hard fescue, 33% slender wheatgrass, and 17% big bluegrass at a rate of 14 kg/ha was recommended (Leege and Godbolt 1985). However, if the objective is to increase shrub density on elk winter range, grass seeding was not recommended because of potential competition with shrub seedlings. If seeding is done, fall burning followed by seeding was suggested. Seeded species did not become productive until the third growing season.

Perhaps the most comprehensive information derived from this series of investigations was the evaluation of use of prescribed burns by elk. An evaluation of the burns on Sherman Creek was carried out for eight years (Leege 1978). Table 10-8 shows data for Scouler willow and redstem for unburned, fall burned and spring burned sites. Use of redstem was high on all sites, being significantly higher on both burned sites during the light winter of 1971-72 but was rather uniformly high on all sites at other years, reflecting the palatability of this species. Scouler willow was browsed to greater diameters and higher proportions on burned sites generally throughout the study period (Table 10-8). A total of 175 plots of ten square feet in size were evaluated. The study was discontinued before the objective of determining how long the difference in use of burned sites over controls continues, but the differences lasted for at least 8 years. The heavy browsing of Scouler willow may help to keep it within reach longer than lighter browsing, and levels observed on redstem were not severe enough to cause damage, assuming 75% utilization or less is satisfactory.

Table 10-7. Responses of herbs and grasses to prescribed burning in Noseeum Creek 1974-1978 (Leege and Godbolt 1985).

Herbs	Response after burning	Response after seed and burn	Grasses & grasslikes	Response after burning	Response after seed and burn
Common yarrow	no change	no change	Slender wheatgrass	-	present at year 4
Piper's anemone	increase thru 4th yr		Wheatgrass spp.	present 1st and 2d year	
Ballhead sandwort	no change	no change	Canada brome	decrease	decrease
Bluebell bellflower	no change	no change	Columbia brome	no change	no change
Thistles.	no change	no change	Sedges	increase	increase
Pinkfairies	no change	no change	Orchardgrass	not present	present at year 4
Blue-eyed mary	no change	no change	Blue wildrye	decrease	decrease
Tiny trumpet.	no change	no change	Idaho fescue.	no change	no change
Idaho goldthread	no change	no change	Western fescue	decrease	decrease
Fireweed	no change	no change	Sheep fescue	not present	absent at year 4
Willowherb	no change	no change	Panic grass	not present	absent at year 4
Yellow avalanche lily	no change	no change	Bluegrass	no change	no change
Woodland strawberry	decline, then increase		Sandberg bluegrass	not present	absent at year 4

Fragrant bedstraw	decrease	decrease	Bulbous bluegrass	not present	absent at year 4
St. John's wort	no change	no change	Kentucky bluegrass	no change	no change
Wild hollyhock	increase	increase			
Sierra pea	no change	no change			
Tarweed	no change	no change			
Sweetclover	not present	disappeared			
Smallflower miterwort	no change	decline			
Evening primrose	increase	increase			
Wilcox's penstemon	increase	increase			
Varileaf phacelia	no change	no change			
Douglas bistort	no change	no change			
Western brackenfern	no change	no change			
Sheep sorrel	increase	not present			
Lilly-of-the-valley	decrease then increase				
Dandelion	no change	no change			
Yellow salsify	no change	no change			
Common mullein	increase	increase			
Violets	no change	no change			

The studies all suggested that prescribed fire could be used to rejuvenate shrub fields with no ecological damage, and that elk tended to select shrubs that resprouted following the fires over shrubs that were not associated with prescribed fires. The problem that became apparent was age-old. By the time shrub fields were being burned, many had succeeded to conifers and opposition to burning "green timber" persisted. In sum, the burning of 5,144 ha over the 1966-1991 period constituted treatment of only 22% of the entire shrub field winter range in the Lochsa, estimated at 23,142 ha by Norberg and Trout (1957).

If the suggestion put forth by Leege (1971) to burn on a rotation of 10-15 years was implemented then between 1,520 and 2,320 ha should be burned annually, except that the portion within the Selway-Bitterroot Wilderness would have to be excluded from a program to purposefully benefit elk and included in a fire management prescription appropriate for wilderness. At the height of the program, over 400 ha were burned in 7 of the 15 years, with 1621 acres (656 ha) being the greatest amount for one year, 1973. The exact consequences of dropping this program for the elk population may not be known for certain, but as the shrub fields continue to succeed to conifers and forage quantity on winter and spring range gradually declines, the gradual decline of the elk population will inevitably follow, and it will remain for those following the population to document just how that process actually takes place. Lack of funding and worries over conifer regeneration were among the reasons that burning was stopped.

Table 10-8. Willow and redstem utilization on burned and burned sites on Sherman Creek over a 6-year period (Leege 1978c).

	SCOULER WILLOW						REDSTEM CEANOTHUS					
	UN-BURNED TWIGS		FALL 66		SPRING '66		UN-BURNED TWIGS		FALL 66		SPRING '66	
	%	DIA.[1]	%	DIA.	%	DIA.	%	DIA	%	DIA.	%	DIA.
1967	65.7	3.7	-	-	93.5	4.5[2]	62.5	3.4	-	-	91.1[2]	3.3
1968	20.5	3.0	75.0[2]	3.8[2]	84.5[2]	3.3[2]	30.0	3.5	65.6	3.1	49.4	3.5
1970	27.3	2.9	38.6	3.5[2]	63.0[2]	3.2	65.5	3.3	81.4	3.7	61.3	3.4
1972	0	-	12.1	2.5	22.0[2]	3.0	1.1	2.8	54.3[2]	3.5	44.0[2]	3.5

[1] Diameter in mm.

[2] Signifies different from unburned at $P=0.05$

CHAPTER ELEVEN

THE HUMAN DIMENSION

Sportsmen and a few others have left a rich history in the *Letters to the Editor* in the Lewiston Tribune. These letters were scanned from the 1970s through the 1990s when the Lochsa elk population was a serious subject of concern to many. That the elk population was dwindling, and the hunting was not as fruitful as it was earlier prompted many to offer an opinion and a criticism of its management. Since the Idaho Department of Fish and Game and the U.S. Forest Service are regulatory agencies as well as management agencies, they received the brunt of the adverse commentary.

An interesting exchange occurred between Jack O'Connor (1902-1978), the well-known sports writer for Outdoor Life, and a citizen from Weippe, Idaho in 1973. The citizen was highly critical of the Idaho Department of Fish and Game, urging the agency to do something more about predators. O'Connor cites his father, also an outspoken journalist, in an exchange (Lewiston Tribune, October 8, 1973) who:

> .
> " ....loathed people who shot of their mouths with a lot of powder but not much lead and after his fourth snort of Kentucky sour mash bourbon he was apt to fix them with his glittering eye and snarl 'Just where in the hell do you tend bar?' "

O'Connor, a resident of Lewiston, Idaho at the time, was defending the agency against the criticism many levied against it for one reason or another, but most often about predator management. However, just as with the sportsmen who wrote the letters, wildlife biologists are not a group of people who share the same opinions, training, and experiences. Biologists who do not have training in grazing influences and plant succession may well see things differently than those who focus on population dynamics and predation. The change from either sex hunts to bull-only hunts that occurred was one example of different interpretations on how to manage the deteriorating elk population.

The person that O'Connor was criticizing had his supporters, just as O'Connor did. And the agency did respond with the prescribed burning program and the changes to more liberal hunting regulations on black bears and cougars. Other letters show that some outdoorsmen can be astute observers and spend enough time in the woods to witness events that enable them to form a useful opinion. While these observations are anecdotal in nature, which makes them unreliable to generalize from, they have value in showing issues that need intensive investigation. This is especially the case when several informed observers see the same issues and form the same opinions.

One problem is that a group of people can organize a campaign of letter writing that supports one point of view. Public comments on a wolf reduction proposal by IDFG resulted in a Defenders of Wildlife campaign to solicit email opposing the proposal that resulted in 38,000 comments that included verbatim or a slightly changed version of a letter sent out as an example (Hinson 2006). It remained for Bill Hall (1937-2018), the editorial page editor for the Lewiston Tribune, to articulate the appropriate view on the relationships between professional wildlife people and lay outdoorsmen in the editorial entitled: *Don't shoot the biologists* (November 29, 1973). The editorial points out that total blind reliance on professional advice would be unwise just as it would be unwise for hunters to place blind faith in "folk biologists" rather than the real ones.

Indeed, wildlife biologists themselves are still considering whether wildlife management is science-based or not (Decker et al. 2018). These writers concluded that professional recommendations to decision makers, even when heavily informed with scientific data, are only a part of how the decisions are formulated. One conclusion is that while the decision makers benefit from science, they must include knowledge of the situation in the decision. Authors conclude that wildlife management can be considered as being science-informed, as distinguished from being science-based.

Sportsmen were arguing for more predator control in the Lochsa long before IDFG did anything of consequence about it. Only after the work initiated in the 1970s which implicated black bears as major predators of elk calves in the Lochsa did the agency institute more liberal hunting seasons on bears and cougars. And it was the sportsmen writing letters to the editor that put pressure on the agency to seriously consider what to do. At first, consideration of the deteriorating habitat situation took more priority in attempts to stop the decline. The reasoning was that habitat loss was the priority to address. Prior to the calf survival study, the finding that black bears were important predators was not well known and certainly not documented. Studies elsewhere in Idaho did not identify the black bear as being a significant predator on elk calves (Beecham and Rohlman 1994).

Calls for more predator control, which would focus primarily on cougars, would not necessarily have been directed at the black bear.

The intervention by Governor Andrus to eliminate hunting of cows and calves illustrates the difficulties in pursuing research programs in controversial situations. If the efforts to eliminate nutrition as a reason for low calf survival had been continued, and either sex hunting had been continued as well, the elk population quite likely would have continued to decline and calf survival would not have increased because of the predation. Just when the management would have changed to bulls-only regulations and liberal seasons on predators would have been instituted is not known but would probably have happened within a year or two of when it did. The calf survival studies were showing the effects of bear predation by the mid-1970s and with the intensive pressures from the interested public, there was justification both in the information being gathered and public sentiment to initiate the changes. Regardless of this speculation, the subsequent regulation changes intended to stop the decline of the Lochsa elk herd did result in an increase from 1973 to1984. However, the expanded harvests on black bears apparently had no effect on predation on elk calves. Whether sportsmen concerns should had been addressed sooner or not remains a source of conjecture.

A controversy over a logging sale in the Pete King drainage occurred in 1978 that was a harbinger of many such occurrences across the western elk ranges. The Pete King experience was intensively reported in the Lewiston Tribune in August of that year. The Walde Timber Sale was set up to use a selection form of cutting to remove dead and dying timber. It was intended to remove 5 million board feet of the estimated 15 million board feet of timber present using overhead cables that would carry logs to ridgetops and eliminate the common "jammer roads" that were typically employed at the time. There had been extensive logging on the Pete King drainage prior to this proposed sale that had increased access, particularly for hunters as well as for timber harvest.

The game manager at Lewiston reported that the Pete King Creek wintering elk population had dropped from 406 in 1957 to 63 in 1975, largely because of increased access along logging roads for hunters. Mort Brigham, one of the most active lay conservationists in central Idaho, was aware of this decline. He appealed the sale through the Forest Service channels eventually to the chief of the agency. With the Idaho Fish & Game Commission and other prominent conservationists also opposing the sale, the chief decided that the elk habitat guidelines that were available were not being followed and forbid the sale.

In retrospect, the reasonably conservative selective cut as was proposed would not have been the most appropriate way to produce shrubs and forbs that would be used by elk and other wildlife. A more appropriate method would have involved group selection of sufficient acreages to make sufficient forage for a reasonably large number of elk. In addition, enough acreage would have to be cut on calving areas to hopefully minimize concentrating elk during calving, thereby making them less vulnerable to predation. Given that forage was known to be declining, such efforts could have proven useful. Access to the drainage would have to be managed as well, and that eventually happened. However, public concerns were influencing the size and kind of cut that was proposed and elk habitat was not a major reason for the project.

Some of the opposition to logging was also attributable to how fish and wildlife interests and timber interests interacted after the timber harvest was dramatically increased on Forest Service lands in the 1960s. Foresters were often arguing that the logging was good for fish and wildlife, and what were then considered the best forestry practices would do no harm. Objections to that idea were voiced, usually to little effect, by both fish and wildlife people, resulting in a buildup of resentment among the various groups within the agency. However, the consequences of increased road building in prime elk habitat plus logging of riparian zones along important trout and salmon streams were hard to ignore. Initial guidelines produced in eastern Oregon for elk (Thomas et al.1979) set out to rectify the problem and reduce logging and road building activities to levels that would not adversely affect fish and wildlife, especially elk. These guidelines were modified to suit other areas, including northern Idaho (Leege 1984). As of this writing, coordination between the state wildlife agencies and the federal land management agencies has increased but is always in need of improvement. More effective public relations programs involving both agencies help. However, some environmental groups challenge every timber sale so the efforts to manipulate forest conditions are made more difficult. Changes in hunting opportunity and elk populations need to be more effectively addressed on federal lands, especially in areas subject to timber management. In many areas, elk populations have shifted to occupy private lands where hunting is either forbidden or highly regulated and away from federal lands where motorized access is a major influence and habitats have deteriorated. Demands for use of the federal lands for all kinds of recreational activities create serious conflicts that managers must address, inevitably resulting in controversy.

A prescribed fire intended to benefit elk habitat created substantial controversy for the Nez Perce National Forest when it burned beyond the boundaries. The project was in the lower Selway River on Rackliff Creek, a tributary to the Selway River near its confluence with the Lochsa, on September 19, 1991. Intended to burn 1000 acres of winter-spring range, the fire eventually burned 4200 acres and was designated a wildfire on September 24 after an unfavorable weather forecast became available. With insufficient personnel available to meet management objectives, heavy smoke travelling west down the Clearwater River canyon, and insufficient funding available without additional help, the wildfire designation could provide more resources. The inadequate weather forecast that did not predict the westerly winds that caused the fire to burn down slope as well as upslope as intended, was the primary reason for the unintended spread of this fire. This created heavy smoke in communities further west and threatened private property along the Selway River.

A comprehensive review of the entire event was prepared by Nez Perce National Forest staff involved in the episode (Lane et al 1991). The review included results of a community meeting on December 12, attended by 21 citizens, 7 Forest Service employees, and 1 from the Idaho Department of Fish and Game. Forty concerns ranging from issues of public safety to better public information were reported in the review. Several attendees also wrote to their congressmen about the fire which precipitated responses from the Washington office of U.S. Forest Service and the office of the Nez Perce National Forest explaining the detailed planning for the prescribed fire.

Advisory Committees

A Citizens Advisory Committee was established in 1995 to advise the IDFG Region 2 office in Lewiston of issues sportsmen and others had with the department. Clearwater Research Incorporated, of Boise, a public opinion research group, produced a report of opinions held by focus groups for wildlife management that were held in Grangeville and Lewiston in July 1995. The two groups had very different opinions of how the Fish and Game Department was viewed, with the Grangeville respondents much more critical of the agency and its policies than the Lewiston respondents. The major habitat-related issue of concern was road closures. Forest succession issues were not addressed by the participants. A need for more public relations was generally supported. Efforts to post information concerning road closures and for more meetings concerning issues in the region have occurred because of this survey. However, comments provided during the focus group meetings hardly mentioned habitat-related issues, suggesting that few in the groups were hunting in the Lochsa, Selway, or North Fork of the Clearwater River.

Collaborative efforts to address the declining elk population in the upper Clearwater started in the1990s with meetings of the various vested interests in the process, initiated by the Idaho Fish & Game Department. Senator Mike Crapo established an effort in the early 2000s that included most of the stakeholders and resulted in a final report in 2004. Participants included sportsmen and outfitters interested in hunting, environmental community representatives, forest products, the Nez Perce Tribe, and the Clearwater County Commissioners.

An analysis of elk habitat in the Lochsa and part of the North Fork Clearwater drainages was published in March 2012 by the Ecosystem Research Group, Missoula, Montana, as part of this initiative. The analysis compared forests age classes from 1850 with current (2010), showing major reduction in seedling-sapling stands from the 1850 estimates. Forage production estimates of around 1000 pounds per acre in these early stands approximate estimates provided by Leege and Schlegel (1973) and Leege (1976). This report concluded that declining forage availability contributes to the declining elk populations.

The Elk Collaborative Report included recommendations that all delegates could reach consensus on. Basically, the recommendations centered around using more prescribed fire, managing wildfires to let them burn as much as possible, harvesting forest products, and mechanical treatment to restore shrub fields, all where legal and possible. Efforts to manage noxious weeds, especially in the Selway River drainage, were recommended.

Recommendations to increase population objectives to at least 35 calves per 100 cows, address excessive harvests of bulls, fully fund the calf survival studies, better manage cougar and black bear harvest using season lengths and improved access, adequately monitor wolf populations, and conduct a wolf/elk/habitat study using controls. A total of 62 recommendations that were approved by all delegates were developed, including 47 dealing with habitat issues such as fire, timber harvest, and weeds. Another 15 recommendations dealt with elk and predator management. These recommendations were applied by the U. S. Forest Service and the Idaho Department of Fish and Game as much as funding would allow in subsequent years. However, none of these recommendations resulted in enough effort to improve elk habitat to make any difference in the trends.

The final report contains an appendix that includes non-consensus recommendations. Given the diversity of interests, involved, it is to be expected that many recommendations that were brought forward could not be agreed upon. The recommendations also show the areas of greatest concern. A total of 31 non-consensus recommendations involved habitat issues, 32 involved predators and population issues. Habitat issues ranged from logging to prescription burning, wildfire management and weed management. Population issues centered around control of the predation on elk and management of hunter harvest, while access issues were related to road closures and access for logging and other activities.

The efforts to bring together as many stakeholders as possible illustrates the problems when so many different points of view are involved. Criticisms of the agencies did result in more effort to inform and involve sportsmen with meetings and commentary in newspapers, probably the major result. Many recommendations all agreed upon would have been useful if enough funds had been available to coordinate treatments of enough habitat and predators to make a difference to the elk population.

CHAPTER TWELVE

FINAL THOUGHTS

People are used to ignoring the processes of habitat change and their influence on ungulates because these processes are often very slow. Indeed, much of the African fauna evolved in habitats that change seasonally due to precipitation or frequent fires, but do not change in plant composition to any appreciable amount. A prime example of this is the Serengeti wildebeest and the associated fauna (Estes 2014). Over evolutionary time, these ecosystems have adapted to the influences of grazing and fire. Reasonably stable habitats likely facilitated the species diversity that evolved on the African continent. The North American plains are another example of how these grasslands evolved with fires, caused both by aboriginals and lightning. The post-fire vegetation was more palatable for bison, elk, and other wildlife in this system. In contrast, the eastern North American deciduous forest evolved in a humid climate where fires were present but less frequent and habitat change was also less frequent. Insects, pathogens, severe storms, and windthrow were primary influences on this highly complex system. These systems may be contrasted with systems where ungulates depend upon habitat that changes from one plant species complex to another entirely different complex over relatively short time spans, such as the moose of the northern hemisphere taiga and elk in the northern Rocky Mountain forests. Counterparts in Eurasia are similar where forests and wildfires have prevailed over evolutionary time.

> "Looking to the future, in view of the needs of elk and the exacting requirements of recreation based on multiple use, the safest course is to model elk management along natural lines, not only to preserve elk as a living animal, but also, so far as is reasonably possible, to preserve its distinct habits as well as its habitat" (Murie 1951).

In the Jackson Hole of Wyoming, where Murie did most of his work on elk, the shrub-steppe habitats these elk depend on do not change much through long time, even following fires. Woody plants such as sagebrush decline following fire in this system, but eventually re-establish. This is contrasted with elk that depend on forest disturbance that replaces forest with forage species, most especially shrubs and forbs. Elk are products of disturbance in much of Idaho north of the Salmon River, northwestern Montana, northeastern Washington, western Alberta, and British Columbia. Wherever shrubs that come in after disturbances constitute an important part of the diet, elk populations will ebb and flow in accordance with that forage base. The ebb and flow are not directly correlated with the habitat change since winter severity, late summer drought, population density related to forage (K), predation, hunter harvest, and age structure of the elk population will all come into play. Ultimately however, the fundamental relationship between the population and its forage base will dominate the interactions. Should we model elk management along natural lines in this kind of country? If so, we will have a long wait for any individual piece of land to alternate between habitat needed for abundant elk and forests that will not provide as much forage and therefore will harbor much fewer elk.

Elk are indigenous to the Lochsa and were never extirpated from the river system. They were present when European man first entered the region and probably colonized the region not long after glacial retreats allowed suitable habitat to exist. Elk numbers have, since their appearance in the drainage, been influenced by habitat conditions following fires of sufficient size to create the shrub fields that provide extensive forage. Significant increases of elk following the extensive wildfires in the first four decades of the last century might not have resulted in such high numbers if hunters had not been occupied with World War II and predators had not been severely reduced. Research and prescribed burning in the 1960s assumed that the observed heavy browsing and the declines in forage availability as shrub field succession progressed, plus a severe winter, underlay elk population declines beginning in the early 1950s. However, predators likely began to increase during WWII as predator control was discontinued.

Predation may have kept elk numbers below the levels that would have existed if forage availability was the only limiting factor. Evidence for this view occurs twice during the last century. The Lieberg surveys show that extensive tracts of land in the upper Clearwater drainage were regenerating from wildfires and while elk populations appeared to increase in the latter part of the 1800s, no irruptive behavior was noted. While inaccessibility of the Lochsa River drainage during the significant wildfire years between 1900 and 1934 would have limited human exploitation, and predation may have constrained an elk population response initially, eventually the huge proliferation of forage overrode the factors that limited the population. Limited access until the construction of Highway 12 undoubtedly restricted harvests until the 1960s. Population reductions following the opening of the highway fostered the change to bulls-only harvest north of the river in 1975 and in the entire drainage in 1976 which resulted in temporary increases in elk until declines in forage availability and possibly increased predation again became prevalent. Increased hunter access via well-developed road systems and habitat alteration that fragmented elk security cover further reduced numbers. After cow and calf hunting was curtailed, survival of bulls was reduced, especially in areas of extensive access. Proliferation of predators, most especially gray wolf and possibly black bear, also suppressed elk populations in recent time. Predation undoubtedly accelerated the current decline of the Lochsa elk population. The severe winter conditions of 1996-97 reduced the elk population to levels where predation became a limiting factor. Also, winter mortality coincided with the introduction of wolves. Prior to the 1996-97 winter, the elk population had increased regardless of bear and mountain lion predation to levels that quite likely exceeded the forage-based habitat capacity.

The influence of aboriginal hunting on this elk population is unknown because harvest is not reported by the tribes. However, it is unlikely that aboriginal take was high enough to influence population trends, even if it might have fluctuated between small harvests to harvests large enough to either temporarily alter elk distributions or reduce numbers in local areas.

Elk in the Lochsa will undoubtedly continue to be present, but in much more limited numbers than occurred in the latter half of the 20th century. While hunting is not a growth sport in the United States, demand for elk hunting in the western states is virtually insatiable. Nevertheless, the increased attention that a broader array of resources on national forests receive by both society and land managers means that elk will be viewed as an important component of a resource complex that must be managed within the context of other values which include timber, recreation, other wildlife, fish, and the watershed (Quigley and Cole 1997).

If past is prologue for the future, then we grant that man's activities will have major influence on elk inhabiting the Lochsa, but we must also recognize that forces in nature over which we have little control will also have influence. Man's activities include not intervening, as when fires are not allowed to burn, as well as active intervention as when prescription fire or predator control is practiced. There is substantial uncertainty in what the future will bring, as climatic patterns change, and forest conditions also change. The inexorable progression of the all-important shrub fields towards conifer forest will continue for the present, but eventually wildfire or pathogens will almost certainly reclaim these forests barring substantial intervention. Wellner (1970) pointed out that the forests of the northern Rocky Mountains virtually all originated from fires prior to man's intervention and there is no reason to think that the future will be different except where logging occurs. It is highly questionable as to whether we can or even should attempt to prevent wildfire in forests such as in the upper Clearwater Basin when climate conditions favoring these fires occur, as the 1988 fire storms across the western United States illustrate. Brown et al. (1994) concluded that the area in the Selway-Bitterroot Wilderness burned by stand replacement fires was 1.5 times greater during the presettlement period than during the 1979-1990 period. Efforts to return the Selway-Bitterroot Wilderness to a more natural condition using natural fires should increase productive elk habitat if the proliferation of exotic weeds can be effectively addressed. How elk respond will depend upon predator levels as well.

If current trends prevail over time, the elk population in the Lochsa will stabilize at some lower unspecified level that the limited forage resources can support and predation pressures allow. This level will be considerably below that level that provided so much hunting opportunity in past times. Further, while we may decide to intervene we may be unable to affect enough area and control enough predators to make a difference. If we wish to attempt an intervention then we must deal with the knowledge that our abilities to achieve a sustainable program of habitat improvement using prescribed fire failed for numerous reasons, not among the least being lack of financial support and our inability to affect enough area to maintain shrub fields of sufficient size to sustain elk populations. Efforts to manage predators did not have much effect either, even if the research on Coolwater Ridge demonstrated that intensive removals of bears would increase calf survival. The impact of black bear predation can be reduced by increasing the cow segment of the elk population, not through bear removal. It's only possible for bears to take a certain number of calves during the 2-3-week period of vulnerability. As more calves are born, the significance of bear predation diminishes. Wolves, on the other hand, have a more significant impact from a population perspective.

Many elk populations in the north Idaho region occur in forest interspersed with agricultural areas, including croplands and grazing pastures. These populations can be managed at levels that landowners tolerate. These kinds of situations are made more difficult where private ownership prohibits hunting in sufficient areas to reduce harvests. This is a different situation than areas of private and commercial forest lands that are not associated with agricultural activities. We can consider the Lochsa population as being a subpopulation of a metapopulation that occupies similar habitat in a broader region. We would then consider habitat conditions in the entire region in Idaho north of the Salmon River. Today, thrifty elk populations occur in north Idaho where logging has created the seral shrub fields they depend upon. We could manage elk inhabiting this broad region which has similar transition-type range. This means crossing national forest boundaries and other public lands, industrial forests and other private timberlands. It also crosses two and perhaps three different regions of the Idaho Department of Fish and Game. Elk habitat management could be addressed at this broad landscape level with planning to predict where seral shrub fields will occur in the future. Planning that incorporates projections for timber harvest on both industrial and other private forests as well as the public lands would be involved. Wildfire management would be included in this kind of large-scale management. The Lochsa elk population would then fluctuate contingent upon the natural forces affecting the habitat and we could broaden our efforts to understand where timber management can be extensive and flexible enough to create shrub habitat. Population management will have to account for habitat change and increased access as well as private holding where no hunting is allowed. Some combination of prescription fire, prescribed wild fire, access management, weed management, predator management, and treatment of second-growth forests might be implemented on a scale that would insure a huntable population of elk north of the Salmon River in Idaho. Many second-growths stands that are replacing the shrub fields are now or shortly will be merchantable and thus if logged would develop shrub fields.

Prescribed fire could be used to regenerate shrub communities from second-growth conifer stands on winter ranges where timber harvest is not an option. In the Lochsa, burning steep south-facing slopes which are sparsely covered with vegetation and are not likely to respond with much shrub and understory production is of less value than creating areas that can be used at other times of the year as well. Shrub fields on productive sites which have grown out-of-reach to elk may also be considered for burning. The extensive logging in the upper portions of the Lochsa did produce quantities of forage for elk, but their use was limited to the summer months because of deep snows and predators and did not appreciably affect population trends. We can be relatively certain that burned areas which exhibit extensive stands of resprouting and new shrubs will draw elk at least to the peripheries, as well as their predators. These situations should be monitored, and population responses should be evaluated. Predator management would have to be integrated into these situations as well.

While management of habitat currently involves extensive uncertainty, the large amount of information that is available on plant community responses to wild and prescribed fire in this region does provide direction in management. Location, timing, intensity, and amount of burning to be accomplished have been identified and recommended. If we try to guide ecosystems towards some range of conditions that we agree are suitable, then the information may be used to develop predictions about plant community response which can be verified through monitoring (Murphy and Noon 1991). The work on stand prognosis models provides a basis for developing predictions.

Options in management of the Lochsa population itself are limited because of the deteriorating habitat. Elk harvests have been modified by changing season opening dates and lengths and limiting harvest to bulls. Emphasis on managing predation to improve calf survival is ongoing. Efforts to improve bull survival also has justification for biological and social reasons. Work in Oregon (Noyes et al. 1996, Trainer and Lightfoot 1970), Utah (Prothero et al. 1979), and Colorado (White et al. 2001) implies that the presence of substantial numbers of mature bulls in elk populations facilitates calf production and synchrony in parturition which can increase survival. Hunters and recreationists also benefit from the presence of more mature bulls in populations. Efforts to increase bull survival may lead to more restrictive hunting seasons, possibly to limited entry hunts which specify antler restrictions. However, as hunter success declines and the number of elk seen decreases, hunter participation appears to self-regulate to some degree.

Elimination of either-sex hunting in the mid-1970s in back country areas as well as other areas in the Clearwater River drainage contributed to substantial increases in elk, illustrating the ability of the hunter to harvest substantial numbers of animals effectively regardless of amount of access. This experience also demonstrates that hunter harvest can be a major limiting factor in the Lochsa. Access management will always be a major factor in managing take.

Harvests of cows to reduce pressures on the forage base to improve calf survival are unlikely because of the predation. While predators had been suspected of reducing survival of elk calves and older age classes in the upper Clearwater since the 1930s, it wasn't until the Coolwater Ridge studies demonstrated that the black bear was the major predator on elk calves and that bears could be important predators (Schlegel 1986). This was subsequently corroborated when the Yellowstone grizzly was confirmed as a major calf predator (Singer et al. 1997). Bears are well known to be omnivorous, and the food groups that Craighead et al. (1982) describe show how grizzly food habits change as different plants and animals become available. This description applies to black bears as well (Beecham and Rohlman 1984). Experience in the Lochsa shows that higher cow populations, experimental removal of bears, and heavy bear harvest by hunters in spring can reduce predation enough to increase calf survival.

However, the longer-term record shows that even liberal general season regulations have not decreased bear populations nor increased calf survival. Couple this with the declining habitat conditions means that predation should probably be viewed as accelerating the decline in elk populations over what would have happened in the absence of predation. The recent fires on the Powell District that have burned over 160,000 acres and have not caused the elk population to increase have to be related to predation as well as continued conifer regeneration on spring-winter ranges at lower elevations.

Adaptive management is a concept that was developed by British Columbia researchers in the 1980s (Walters 1986) and has been expanded upon in a variety of contexts since. A basic tenet of adaptive management is that we will treat management as an experiment, establish testable objectives which may be treated as hypotheses, carry out the necessary perturbation of the system, and monitor and modify as the information dictates. Management by experiment (MacNab 1983, Sinclair 1991) has been recognized as a progressive and defensible means to address uncertainties involved in wildlife management problems. Population management that includes hunting may be addressed as an experiment. Habitat management which may include use of wild and prescribed fire or logging and associated road management may also be designed as an adaptive management process. In fact, given that there is considerable uncertainty and surprises involved in management activities of all kinds, whether we view the management process as an experiment or not, it often is anyway.

The real lesson that we learn from the Lochsa elk history is that forces in nature are largely responsible for the major trends in population and habitat. The corollary is that man's efforts to intervene did not appreciably alter these major trends. Whether we intervene more effectively in the future, or whether we even should, are questions for society and managers to address in this century.

LITERATURE CITED

Alldredge, M.W., J.M. Peek, and W.A. Wall. 2001. Shrub community development and annual productivity trends over a 100-year period on an industrial forest of northern Idaho. Forest Ecology and Management 152: 259-273.

Alldredge, M.W., J.M. Peek, and W.A. Wall. 2002. Nutritional quality of forages used by elk in northern Idaho. Journal of Range Management 55: 253-259.

Arno, S.F. 1980. Forest fire history in the northern Rockies. Journal of Forestry 78: 460-465.

Asherin. D.A. 1973. Prescribed burning effects on nutrition, production and big game use of key northern Idaho browse species. Univ. Idaho PhD thesis. 96 pp.

Ausband, D. E., L. N. Rich, E. M. Glenn, M. S. Mitchell, P. Zager, D. A.W. Miller, L. P. Waits, B. B. Ackerman, and C. M. Mack. 2014. Monitoring gray wolf populations using multiple survey methods. Journal of Wildlife Management 78:335-346.

Ausband, D.E. 2016. Gray wolf harvest in Idaho. Wildlife Society Bulletin 40:500-505.

Aycrigg, J.L, and E.O. Garton. 2014. Linking metapopulation structure to elk population management in Idaho: a genetic approach. Journal of Mammalogy 95:59-614.

Baird, L., and Baird, D. 2003. In Nez Perce Country: accounts of the Bitterroots and the Clearwater after Lewis and Clark. University of Idaho Library, Moscow, Idaho. 288 pp.

Barrett, S.W., S.F. Arno,and J.P. Menakis. 1997. Fire episodes in the Inland Northwest (1540-1940) based on fire history data. U.S. Forest Service General Technical Report INT-GTR-370. 17 pp.

Beecham, J.J. and J. Rohlman. 1994. A shadow in the Forest: Idaho's black bear. University of Idaho Press. Moscow, Idaho. 245 pp.

Bergerud, A.T., and J.P. Elliott. 1986. Dynamics of caribou and wolves in northern British Columbia. Canadian Journal of Zoology 64:1515-1529.

Bethlamy, N. 1967. Effect of exposure and logging on runoff and erosion. USDA Forest Service Research Note INT-61.

Boyce, M.S. 1989. The Jackson Hole elk herd. Cambridge University Press, Cambridge. 306pp.

Boyce, M.S. 2018. Wolves for Yellowstone: dynamics in time and space. Journal of Mammalogy 99: 1-11.
Boyd, R.J. 1970. Elk of the White River Plateau, Colorado. Colorado Game, Fish and Parks Technical Bulletin 25. 124pp.
Brown, J.K., S.F. Arno, S.W. Barrett, and J. P. Menakis. 1994. Comparing the prescribed natural fire program with presettlement fires in the Selway-Bitterroot Wilderness. International Journal of Wildland Fire 4: 157-168.
Brunelle, A. et al. 2005. Holocene fire and vegetation along environmental gradients in the Northern Rocky Mountains. Quarternary Science Reviews 24:2281-2300.
Christensen, A.G., L.J. Lyon and T.N. Lonner, compilers. 1991. Proceedings Elk Vulnerability Symposium, Montana State University, Bozeman, Montana. 330pp.
Clearwater National Forest 1948. Report of wildlife survey. Orofino, Idaho.
Clutton-Brock, T.H. 1984. Reproductive effort and terminal investment in iteroparous animals. American Naturalist 123:212-229.
Clutton-Brock, T.H., F.E.Guinness and S.D. Albon. 1982. Red deer behavior and ecology of two sexes. University of Chicago Press, Chicago, Illinois. 378 pp.
Clutton-Brock, T.H, S.D. Albon and F.E. Guinness. 1987. Interactions between population density and maternal charactistics affecting fecundity and juvenile survival in red deer. Journal of Animal Ecology 56:857-871.
Cook, J.G., L. J. Quinlan, L. L. Irwin, L. D. Bryant, R. A. Riggs and J. W. Thomas. 1996. Nutrition-Growth relations of elk calves during late summer and fall. Journal of Wildlife Management 60: 528-541.
Cook, R.C., J. G. Cook, D. L. Murray, P. Zager, B. K. Johnson,and M. W. Gratson.2001. Development of predictive models of nutritional condition for Rocky Mountain elk. Journal of Wildlife Management 65: 973-987.
Cook, R.C. et al. 2001. Nutritional influences on breeding dynamics of elk. Canadian Journal of Zoology 84:845-853.
Cook, J.G. et al. 2004. Effects of summer –autumnal nutrition and parturition date on reproduction and survival of elk. Wildlife Monographs 155.
Cook, J. G., L.L. Irwin, L.D. Bryant, R.A. Riggs and J.W. Thomas. 1998. Relations of forest cover and condition of elk: A test of the thermal cover hypothesis in summer and winter. Wildlife Monographs 141. 61pp.

Coster, R.A. and J.A. Parsell. 1936. Nez Perce winter game studies report and recommendations. Unpublished. report on file. Nezperce National Forest, Grangeville, ID. 40 pp.

Coughenour, M.B. and F.J. Singer. 1996. Elk population processes in Yellowstone National Park, under the policy of natural regulation. Ecological Applications 6:573-593.

Craighead, J.J., J.S. Sumner, and G.B, Scaggs. 1982. A definitive system for analysis of grizzly bear habitat and other wilderness resources. Wildlife-Wildlands Institute Monograph 1. Missoula, Montana. 279 pp.

Darling, F.F. 1937. A herd of red deer. Oxford University Press, London. 215 pp.

Decker, D.J., J.F. Organ, A.B. Forschen, M.V. Schiavone, and A.K. Fuller. 2018. Wildlife management is science based: myth or reality? The Wildlife Professional 12: 30-32. The Wildlife Society, Bethesda, MD.

Egan, T. 2009. The big burn. Houghton Mifflin Harcourt Publishing Company. New York, New York. 324 pp.

Eddy, L.E. and E.A. Rauma. 1936. Fish and Game Clearwater Winter Game Study. Unpublished report on file at Clearwater National Forest, Orofino, Idaho. May 14, 1936. 14pp.

Estes, R.D. 2014. The gnu's world. University of California Press. 368 pp.

Dalke, P.D., R.D. Beeman, F.J.Kindel, R.J. Robel, and T.R. Williams. 1965. Use of salt by elk in Idaho. Journal of Wildlife Management 29:319-332.

Ferguson, D.E. and C.E. Carlson. 1993. Predicting regeneration establishment with the prognosis model. U.S. Department of Agriculture Forest Service Research Paper INT-467. 54 pp.

Finklin, A. J. 1983. Weather and climate of the Selway-Bitterroot Wilderness. University of Idaho Press. 144 pp.

Follis, T.B. and J.J. Spillett. 1972. A new method for rumen sampling. Journal of Wildlife Management 36:1336-1340.

Freddy, D.J. 1987. The White River elk herd: a perspective 1960-1985. Colorado Division of Wildlife Technical Publication 37. 64pp.

Fry, W.E. and J.A. Parsell. 1937. Report on winter game studies, Lochsa part of Nezperce and Clearwater National Forests. Report on file, Nezperce National Forest, Grangeville, ID. 36 pp plus tables.

Fry, W.E. and A. J. Parsell [J. A. Parsell]. Report on winter Studies Lochsa River, Lower Selway River, and Meadow Creek. Lochsa Part of Nez Perce and Clearwater National Forests. Area November 15, 1936-April 15, 1937. Report on file, Nezperce National Forest, Grangeville.

Fryxell, J.M., and A.R.E. Sinclair. 1988. Causes and consequences of migration by large herbivores. Trends in Ecology and Evolution. 3:237-241.

Gaillard, J-M., M. Fest-Bianchet, and N.G. Yoccoz. 1998. Population dynamics of large herbivores: variable recruitment with constant adult survival. Trends in Ecology and Evolution 13:58-63.

Gaillard, J.M. et al. 2000. Temporal variation in fitness components and population dynamics of large herbivores. Annual Review of Ecology and Systematics 31: 367-393.

Garrott, R.A., P.J. White, and J.J. Rotella. 2009. The Madison headwaters elk herd: stability in an inherently variable environment. pp191-216 *in* R.A. Garrott, P.J. White and F. Watson, editors. The ecology of large mammals in central Yellowstone. Elsevier, Inc. San Diego, California. 693 pp.

Garshelis, D.L. 1994. Density-dependent population regulation of black bears. Pp 3-14 in M. Taylor, ed. Density-dependent population in black bears. International Conference on Bear Research and Management Monograph Series 3: 43pp.

Gasaway, W.C., R.O Stephenson, J.L. Davis, P.E.K Shepherd, and O.E. Burris. 1983. Interrelationships of wolves, prey, and man in interior Alaska. Wildlife Monographs 84. 50pp.

Gasaway, W.C., R.D. Boertge, D.V. Grangaard, D.G. Kelleyhouse, R.O. Stephenson, and D. G. Larsen. 1992. The role of predation in limiting moose low densities in Alaska and Yukon and implications for conservation. Wildlife Monographs 120.

Geist, V. 1971. The relation of social evolution and dispersal in ungulates with special reference to old world deer and the genus Bison. Quaternary Research 1:283-315.

Geist, V. 1998. Deer of the World, their evolution, behavior, and ecology. Stackpole Books., Mechanicsburg, PA. 421 pp.

Gower, C.N., R.A. Garrott, P.J. White, S. Cherry, and N. G. Yoccoz. 2009. Elk group size and wolf predation: a flexible strategy when faced with variable risk. Pp401-422 *in* R.A. Garrott, P.J. White and F. Watson, editors. The ecology of large mammals in central Yellowstone. Elsevier, Inc. San Diego, California. 693 pp.

Gratkowski, H. 1962. Pregermination treatments for redstem ceanothus seeds. USDA Forest Service Research Paper PNW 156. 10 pp.

Green, P. and D. Talbert. 1991. Soil and vegetation response to prescribed burning for winter range enhancement. Report on file, Nez Perce National Forest, Grangeville, Idaho. 11 pp.

Greenwood, W.R. and D.A. Morrison. 1967. Reconnaissance geology of the Selway-Bitterroot Wilderness Area. Idaho Bureau of Mines and Geology Information Circular 18. 16pp.
Greer, K. R. 1966. Fertility rates of the northern Yellowstone elk populations. Western Association of State Fish and Game Commissioners 46: 123-128.
Griffin, K. A. et al. 2011. Neonatal mortality of elk driven by climate, predator phenology and predator community composition. Journal of Animal Ecology 80:1246-1257.
Guthrie, R.D. 1966. The extinct wapiti of Alaska and Yukon Territory. Canadian Journal of Zoology 44:47-57.
—. 1968. Paleoecology of the large-mammal community in interior Alaska during the Pleistocene. American Midland Naturalist 79:346-363.
Hadly, E. 1990. Late holocene mammalian fauna of Lamar Cave and its implications for ecosystem dynamics in Yellowstone National Park, Wyoming. MS. Thesis, Northern Arizona University. 128 pp.
Hamilton, L. 1997. Snowbound. Washington State University Press, Pullman, Washington. 228 pp.
Harrison, J.L. 1958. Range and movements of some Malayan rats. Journal of Mammalogy 38: 190-206
Hash, H.S. 1973. Movements and food habits of the Lochsa elk. M.S. Thesis, Univ. Idaho, Moscow. 76pp.
Hatter, J. 1949. The status of moose in North America. Transactions North American Wildlife Conference. 14: 492-501.
Hayden, J. 2014. Idaho Department of Fish and Game surveys and Inventories. FY 2014 Statewide Report. Mountain lion Study 1, Job 8. Boise, Idaho.13-14.
Hayden, J. 2017. Wolf surveys and inventories. Idaho Department of Fish and Game. Boise. 20pp.
Hayden, J. 2018. Black bear surveys and inventories. Idaho Department of Fish and Game. Boise. 71pp.
Hayne, D.W. 1949. Calculation of size of home range. Journal of Mammalogy 30:1-18.
Hebblewhite, M. and D.H. Pletscher. 2002. Effects of elk group size on predation by wolves. Canadian Journal of Zoology 80: 800-809.
Hebblewhite, M., and E.H. Merrill. 2009. Tradeoffs between predation risk and forage differ between migrant strategies in a migratory ungulate. Ecology 90: 3445-3454.
Hebblewhite, M., and E.H. Merrill. 2011. Demographic balancing of migrant and resident elk in a partially migratory population through forage-predation tradeoffs. Oikos 120:1860-1870.

Hickey, W.O., and T.A. Leege. 1970. Ecology and management of redstem ceanothus, a review. Wildlife Bulletin No. 4, Idaho Dept. of Fish and Game. 18pp.

Hinson, J. 2006. Idaho Department of Fish and Game proposal to reduce wolf numbers in the Lolo Elk Management Zone. Analysis of Public Comments. Northwest Natural Resource Group-LLC. 13pp.

Hobbs, N.T. 1996. Modification of ecosystems by ungulates. Journal of Wildlife Management 60:695-713.

Hooker, L.L. 1972. Effects of prescribed burning on soils and vegetation in the Lochsa River region of northern Idaho. M.S. Thesis, Univ. Idaho, Moscow. 75pp.

Hornocker, M.G. and S. Negri. Eds. 2010. Cougar ecology and conservation. University of Chicago Press, Ltd., London. 306 pp.

Houston, D.B. 1982. The northern Yellowstone elk: ecology and management. MacMillan Publ. Co., Inc., New York. 474pp.

Irwin, L.L. and J.M. Peek. 1983. Elk, *Cervus elaphus*, foraging related to forest management and succession. Canadian Field-Naturalist 97:443-447.

Johnson, D.E. 1951. Biology of the elk calf, *Cervus canadensis nelsoni*. Journal of Wildlife Management 15: 396-410.

Jones, J.K., Jr., D.C. Carter, H.H. Genoways, R.S. Hoffmann, and D.W. Rice. 1982. Revised checklist of North American mammals north of Mexico, 1982. Occasional Papers of the Museum, Texas Tech University, No. 80. Lubbock, Texas.

Judd, S.L. 1971. Movements and migration of the Lochsa elk. M.S. Thesis, Univ. Idaho, Moscow. 39pp.

Kindell, F. J. 1958. Salt in the management of elk and other wildlife in the vicinity of Coolwater Ridge area of Idaho. Idaho Fish and Game Department Wildlife Restoration Division Project W-85-R-8. Job No. 7. 69 pp.

Koch, E. 1935. The passing of the Lolo Trail. Journal of Forestry 33: 98-104.

—. 1998. Forty years a forester. Mountain Press Publishing Company. Missoula, Montana. 206 pp.

Kurten, B., and E. Anderson. 1980. Pleistocene mammals of North America. Columbia University Press. New York. 442 pp.

Landall, L. and J.C. Fager. 1935. Report on winter game studies conducted in 1934-1935 in the Selway and Lochsa Drainages. Unpublished report on file, Nez Perce National Forest, Grangeville, Idaho. 65pp.

Lane,,C. et al. 1991. Rackliff Fire Review. Report on file, Nez Perce National Forest, Kamiah, Idaho. 24 pp plus appendices

Leege, T.A. 1968. Prescribed burning for elk in northern Idaho. Proc. Tall Timbers Fire Ecology Conference 8:235-253.

—. 1969. Burning seral brush ranges for big game in northern Idaho. Proceedings North American Wildlife and Natural Resources Conference 34:429-438.

—. 1971. Changing vegetation means fewer elk. Idaho Wildlife Review 24:7-9.

—. 1978a. Changes in browse intercept, production, and seedlings after burning-Holly Creek. Final Report W-160-R- Study 1 Job 2. Idaho Department of Fish & Game, Boise. 11pp.

—. 1978b. Changes in browse production after browsing on four cardinal aspects-Polar Ridge. Final Report W=160-R Study 2, Job 3. Idaho Department of Fish& Game, Boise 19 pp.

—. 1978c. Changes in browse utilization after burning. Final Report. Project W-160-R, Study II, Job 2. Idaho Department of fish and Game. Boise. 5 pp.

—. 1979. The movements and habitat preferences of elk transplanted into the Pete King drainage. Idaho Department of Fish & Game Project W160-R-3-5.

—. 1979. Effects of repeated prescribed burns on northern Idaho elk browse. Northwest Science 53:107-113.

—. 1984. Guidelines for evaluating and managing summer elk habitat in northern Idaho. Idaho Dep. Fish and Game Wildlife Bulletin 11. 37pp

Leege, T.A., and W.O. Hickey. 1971. Sprouting of northern Idaho shrubs after prescribed burning. Journal of Wildlife Management 35:508-515.

Leege, T.A., and M.C. Fultz. 1972. Aerial ignition of Idaho elk ranges. Journal of Wildlife Management 36:1332-1336.

Leege, T.A., and W.O. Hickey. 1975. Growth and dieback of redstem in Idaho. Northwest Science 49:58-64.

Leege, T.A., and W.O. Hickey. 1977. Elk-snow-habitat relationships in the Pete King drainage, Idaho. Wildlife Bulletin No. 6, Idaho Dept. of Fish and Game. 23pp.

Leege, T.A., and G. Godbolt. 1985. Herbaceous response following prescribed burning and seeding of elk range in Idaho. Northwest Science 59:134-143.

Leege, T.A., and M.W. Schlegel. 1973. Elk ecology. Job progress report, project W-160-R-2. Idaho Fish and Game Dept., Boise, 38pp.

Lieberg, J.B. 1899. Bitterroot Forest Reserve. pp 253-282 in 19th Annual Report, US Geological Survey Part V. Forest Reserves. 400pp.

Lovaas, A.L. 1978. People and the Gallatin elk herd. Montana Fish and Game Dept., Helena. 44pp.

Lukacs, P.M. et al. 2018. Factors affecting elk recruitment across ecotypes in the western United States. Journal of Wildlife Management 82. http:DOI:10/1002/jwmg.21438.

Lyon, L.J. and W.F. Mueggler.1968. Herbicide treatment of north Idaho browse evaluated six years later. Journal of Wildlife Management 32: 538-541

Macnab, J. 1983. Wildlife management as scientific experimentation. Wildlife Society Bulletin 11: 397-401.

Mao, J.S., M.S.Boyce, D.W.Smith, F.J. Singer, D.Vales, J.M. Vore and E.H.Merrill. 2005. Habitat selection by elk before and after wolf reintroduction in Yellowstone National Park. Journal of Wildlife Management 69: 1691-1707.

McLean, L.S. 1972. Movements of the Lochsa elk herd. M.S. Thesis, Univ. Idaho, Moscow. 65pp

MacNulty, D.R., D.R. Stahler, C. T. Wyman, J. Ruprecht, and D.W. Smith. 2016. The challenge of understanding northern Yellowstone elk dynamics after wolf reintroduction. Yellowstone Science 24 :25-33.

Mehringer, P.J., Jr, S.F. Arno, and K.L. Petersen. 1977. Postglacial history of Lost Trail Pass Bog, Bitterroot Mountains, Montana. Arctic and Alpine Research 9:345-368.

Merrill, E.H., H.F. Mayland, and J.M. Peek. 1982. Shrub responses after fire in an Idaho ponderosa pine community. Journal of Wildlife Management 46: 496-502.

Meske, T.A. 1972. Big game habitat improvement studies in the Dworshak Reservoir area. Idaho Department of Fish and Game. Job Completion Report. Boise: 31pp.

Middleton, A.D., M.J. Kauffman, D.E. McWhirter, J.G. Cook, R.C. Cook, A.A. Nelson, M.D. Jimenez and R.W. Klaver.2013. Animal migration amid shifting patterns of phenology and predation: lessons from a Yellowstone elk herd. Ecology 94: 1245-1256.

Moeur, M. 1985. Cover: a user's guide to the canopy and shrubs extension of the stand prognosis model. U.S. Department of Agriculture Forest Service General Technical Report INT-190. 49 pp.

Monroe, C.L. 1914. After game poachers on the Idaho-Montana line. Idaho Fish & Game 1(1): 22-23. (price of magazine: 10 cents)

Moore, B. 1996. The Lochsa story: land ethics in the Bitterroot Mountains. Mountain Press Publishing Company, Missoula, Montana. 458 pp.

Mueggler, W.F. 1965. Ecology of seral shrub communities in the cedar-hemlock zone of northern Idaho. Ecological Monographs 35: 165-185.

—. 1966. Herbicide treatment of browse on a big game winter range in northern Idaho. Journal of Wildlife Management 30: 141-151.

Murie, O.J.1951. The Elk of North America. Stackpole Co. Harrisburg, PA.
Murphy, D.D. and B.D. Noon. 1991. Coping with uncertainty in wildlife biology. Journal of Wildlife Management. 55:773-782.
Mysterud, A. 2013. Ungulate migration, plant phenology, and large carnivores;' the times they are a-changin' . Ecology 94:1257-1261. 2013
Nelson, L.J and J.M. Peek. 1984. Effect of survival and fecundity on rate of increase of elk. Journal of Wildlife Management 46:535-540.
Norberg, E.R. and L. Trout. 1957. Clearwater game and range survey. Idaho Department of Fish & Game Final Report, Project W-112-R. 240pp.
Nowlin, R.A. 1974. Prescribed burning effects on in vitro digestibility of elk browse. MS Thesis, University of Idaho, Moscow, Idaho. 27pp.
Noyes, J.H. et al. 1996. Effects of bull age on conception dates and pregnancy rates of cow elk. Journal of Wildlife Management 60: 508-516.
O'Gara, B.W. 2002. Taxonomy. pp 3-65 *in* D.E. Toweill and J.W. Thomas compilers and editors. North American Elk: ecology and management. Smithsonian Institution Press, Washington, D.C. 962 pp.
O'Gara, B.W. and R.G. Dundas. 2002. Distribution: Past and Present. pp 67-119 *in* D.E. Toweill and J.W. Thomas compilers and editors. North American Elk: ecology and management. Smithsonian Institution Press, Washington, D.C. 962 pp.
Orme, M.L. and T.A. Leege. 1976. Emergence and survival of redstem (Ceanothus sanguineous) following prescribed burning. Tall Timbers Fire Ecology Conference 14: 391-420
Ozoga, J.J. 1968. Variations in microclimate in a swamp conifer deer yard in northern Michigan. Journal of Wildlife Management 32: 574-585.
Parsell, J. (no date). The elk problem in the Selway. Unpubl. report on file, Nez Perce National Forest, Grangeville, Idaho.
Parsell, N. 1990. Major Fenn's country. Pacific Northwest National Parks and Forests Association, Seattle, Washington. 43 pp.
Peek, J.M., D. Talbert, S. Blair, R. Guse, and C. Hunter. 1992. An assessment of elk pellet group, Scouler willow, and redstem ceanothus distribution on the lower Selway River burns, Summer 1991. Unpublished report on file, Nez Perce National Forest, Selway District, Kooskia, Idaho. 21pp.
Peek, J.M. 2014. Annual changes in bluebunch wheatgrass biomass and nutrients related to climate and wildfire. Northwest Science 88: 129-139.
Peterson, R.L. 1955. North American moose. University of Toronto Press. 280pp.

Perez-Barberia, E.J., R.J. Hooper, and I.J. Gordon.2103. Long-term density-dependent changes in habitat selection in red deer (Cervus elaphus). Oecologia 173: 837-847.

Picton, H.D. and R.R. Knight. 1969. A numerical index of winter conditions of use in big game management. pp 29-38 *in* A.O. Haugen, ed. Proc. Snow and ice in relation to wildlife and recreation symposium. Iowa State Univ., Ames, Iowa.

Picton, H.D. and I.E. Picton. 1975. Saga of the Sun: a history of the Sun River elk herd. Montana Dept. of Fish and Game, Helena. 55pp.

Polziehn, R.O., J. Hamr, F.F. Mallory, and C. Strobeck. 1998. Phylogenetic status of North American wapiti (*Cervus elaphus*) subspecies. Canadian Journal of Zoology 76: 998-1010.

Polziehn, R.O., and C. Strobeck. 1998. Phylogeny of wapiti, red deer, sika deer and other North American cervids as determined from mitochondrial DNA. Molecular Phylogenetics and Evolution 10: 249-258.

Prothero, W.L., J.J. Spillett, and D.F. Balph. 1979. Rutting behavior of yearling and mature bull elk: some implications for open bull hunting. pp 160-165 *in* M.S. Boyce and L.D. Hayden-Wing, eds. North American elk: ecology, behavior and management. Univ. of Wyoming, Laramie.

Pyne, S.J. 2008. Year of the fires. Mountain Press Publishing Company, Missoula, Montana. 322 pp.

Quick, C.R., and A.S. Quick. 1961. Germination of ceanothus seeds. Madrona 16: 23-30.

Quigley, T.M. and H.B. Cole 1997. Highlighted scientific findings of the interior Columbia Basin ecosystem management project. USDA For Serv. Gen. Tech. Rept. PNW-GTR-404. 34pp.

Quimby, D.C., and J.E. Gaab. 1957. Mandibular dentition as an indication of age in Rocky Mountain elk. Journal of Wildlife Management 21:435-451.

Raithel, J.D., M.J. Kauffman, and D.H. Pletscher. 2007. Impact of spatial and temporal variation in calf survival on the growth of elk populations. Journal of Wildlife Management 71:795-803.

Rich, L.N., et al. 2013. Estimating occupancy and predicting numbers of gray wolf packs in Montana using hunter surveys. Journal of Wildlife Management 77:1280-1289.

Ruth, T.K. 2004. Patterns of resource use by cougars and wolves in northwestern Montana and southeastern British Columbia. PhD dissertation, University of Idaho, Moscow, Idaho.

Samuel, M.D., E.O. Garton, M.W. Schlegel, and R.G. Carson.1987. 1987. Visibility bias during aerial surveys of elk in northcentral Idaho. Journal of Wildlife Management 51:622-630.

Sappington, R.L. 1994. The prehistory of the Clearwater River region, North Central Idaho. University of Idaho Anthropological Reports 95. 422 pp.

Sappington, R.L. and C.D. Carley. 1989. Archeological investigations at the Beaver Flat and Pete King Creek sites, Lochsa River, North Central Idaho. University of Idaho Anthropological Reports 89. 127 pp.

Schlegel, M.W. 1976. Factors affecting calf elk survival in north-central Idaho: a progress report. Proc. West. Assoc. State Game and Fish Comm. 56:342-355.

—. 1980. Analysis of bull only hunting in the Lochsa River drainage, Idaho. Proceedings Western Association of Game and Fish Commissioners 60: 118.

—. 1986. Factors affecting calf survival in the Lochsa elk herd. Project No. W-160-R3B, Job No. 3, Idaho Dept. of Fish and Game. 43pp.

Schlegel, M.W., and T.A. Leege. 1972. Movements and harvest of Glover Ridge elk. Idaho Wildlife Review pp 3-6. September-October 1972. Boise, Idaho.

Smith,D.W. et al. 2016.. Wolf restoration in Yellowstone: reintroduction to recovery.Yellowstone Science 24: 4-11.

Sinclair, A.R.E. 1991. Science and the practice of wildlife management. Journal of Wildlife Management 55:767-773.

Singer, F.J., A. Harting, K.K. Symonds, and M.B. Coughenour. 1997. Density dependence, compensation, and environmental effects on elk calf mortality in Yellowstone National Park. Journal of Wildlife Management 61: 12-25.

Space, R.S. 1964. The Clearwater Story. USDA Forest Service, Missoula, MT. 165 pp.

Stage, A.R. 1973. Prognosis model for stand development. U.S. Department of Agriculture Forest Service Research Paper INT-137. 32pp.

Stickney, P.F. 1965. Note on winter crown kill of C*eanothus velutinus.* Proceedings Montana Academy of Science. 25:52-57.

Thomas, J.W., editor. 1979. Wildlife habitats in managed forests of Oregon and Washington. US Department of Agriculture Handbook 553. US Government Printing Office, Washington, D.C. 512 pp.

Thorne, E.T., R.E. Dean, and W.G. Hepworth. 1976. Nutrition during gestation in relation to successful reproduction in elk. Journal of Wildlife Management 40: 330.335.

Trainer, C.E. and W.C. Lightfoot. 1970. Fertility of yearling male Roosevelt elk in Oregon. Proceedings Annual Conference Western Association State Game and Fish Commissioners. 51: 311-317.

Unsworth, J.W. and L. Kuck. 1991. Bull elk vulnerability in the Clearwater drainage of north-central Idaho. Pages 85-88 *in* A.G. Christensen, L.J. Lyon and T.N. Lonner, compilers, Proceedings Elk Vulnerability Symposium, Montana State University, Bozeman. 330pp.

Unsworth, J.W. and T.C. Ferguson. 1991. Elk habitat security characteristics and hunting season mortality rates. Idaho Dep. Fish & Game Job Progress Rept W-160-R 17.p 14.

Unsworth, J.W., L. Kuck, M.D. Scott, and E.O. Garton. 1993. Elk mortality in the Clearwater drainage of northcentral Idaho. Journal of Wildlife Management. 57:495-502.

Unsworth, J. W., J.W., L.Kuck, E.O.Garton and B.R. Butterfield. 1998. Elk habitat selection on the Clearwater National Forest, Idaho. Journal of Wildlife Management 62:1255-1263.

Urness, P. 1989. Why did bison fail west of the Rockies? Utah Science 50:174-180.

Vaughan, H.W., R.R. Knight, and F.W. Frank. 1973. A study of reproduction, disease and physiological blood and serum values in Idaho elk. Journal of Wildlife Diseases 9: 296-301.

Walters, C.J. 1986. Adaptive management of renewable resources. Macmillan Publishing Company, New York. 363 pp.

Wellner, C.A. 1970. Fire history in the northern Rocky Mountains. Pages 42-64 in The role of fire in the intermountain west. Intermountain Fire Research Council, Missoula, Montana.

Weyermann, G. F., A. Nousianen, and V.L. Erickson. 1950. Winter browse range survey, Lochsa sub-unit. Clearwater National Forest. 8pp.

White, G.C., D.J. Freddy, and J.H. Ellenberg. 2001. Effect of adult sex ratio on mule deer and elk productivity in Colorado. Journal of Wildlife Management 65: 543-551.

White, C.G., P. Zager, and M.W. Gratson. 2010. Influence of predator harvest, biological factors, and landscape on elk calf survival in Idaho. Journal of Wildlife Management 74: 355-369.

Williams, T.R. 1962. The significance of salt and natural licks in elk management. M.S. Thesis, University of Idaho, Moscow. 200pp.

Woodburne, M.O. 2004. Late Cretaceous and Cenozoic mammals of North America. Columbia University Press, New York. 391 pp.

Young, V.A. and W.L. Robinette. 1939. A study of the range habits of elk on the Selway Game Preserve. Univ. Idaho School of Forestry Bulletin 9. 47 pp.

Young, V.A. and G.F. Payne. 1948. Utilization of key browse species in relation to proper grazing practices in cutover western white pine lands in northern Idaho. Journal of Forestry. 46: 35-40.

Zager, P. and J. Beecham. 2004. The role of black bears and brown bears as predators on ungulates in North America. Ursus 17: 95-108.

Zager, P., C. White, and G. Pauley. 2006. Idaho Department of Fish & Game Project W-160-R-33. Draft completion report, elk ecology: factors influencing elk calf recruitment. Idaho Fish & Game, Boise. 44 pp.

APPENDIX

COMMON AND SCIENTIFIC NAMES OF PLANTS USING U.S. DEPARTMENT OF AGRICULTURE NATURAL RESOURCES CONSERVATION SERVICES, 2020. THE PLANTS DATABASE
(http://plants.usda.gov)

	Scientific name	Common Name
Trees	Abies grandis	Grand Fir
	Abies lasiocarpa	Subalpine fir
	Picea engelmannii	Englemann spruce
	Pinus contorta	Lodgepole pine
	Pinus monticola	White pine
	Pinus ponderosa	Ponderosa pine
	Pseudotsuga menziesii	Douglas fir
	Thuja plicata	Western red cedar
	Tsuga heterophylla	Western hemlock
	Tsuga mertensiana	Mountain hemlock
Shrubs	Acer glabrum	Rocky Mountain maple
	Amelanchier alnifolia	Serviceberry
	Ceanothus sanguineus	Redstem ceanothus
	Ceanothus velutinus	Snowbrush ceanothus
	Holodiscus discolor	Oceanspray
	Lonicera utahensis	Utah honeysuckle
	Pachistima myrsinites	Pachistima
	Physocarpus malvaceous	Mallow ninebark
	Prunus emarginata	Bitter cherry
	Salix scouleriana	Scouler willow
	Sambucus racemosa	Red elderberry
	Sorbus sitchensis	Western mountain ash
	Symphoricarpos albus	Common snowberry
Herbs	Achillea millifolium	Common yarrow
(forbs)	Anemone piperi	Piper's anemone

Scientific name	Common Name
Arenaria congesta	Ballhead sandwort
Campanula rotundifolia	Bluebell bellflower
Chamerion angustifolium	Fireweed
Cirsium spp.	Thistles
Clarkia pulchella	Pinkfairies
Collinsia parviflora	Blue-eyed mary
Collomia linearis	Tiny trumpet
Coptis occidentalis	Idaho goldthread
Epilobium angustifolium	Willowherb
Erythronium grandiflorum	Yellow avalanche lily
Fragaria vesca	Woodland strawberry
Gallium triflorum	Fragrant bedstraw
Hypericum perforatum	St. John's wort
Iliamna rivularis	Wild hollyhock
Lathyrus nevadensis	Sierra pea
Madia spp	Tarweed
Maianthemum racemosum	Lilly-of-the-valley
Melilotus officinalis	Sweetclover
Mitella stauropetala	Smallflower miterwort
Oenothera caespitosa	Evening primrose
Penstemon wilcoxii	Wilcox's penstemon
Phacelia heterophylla	Varileaf phacelia
Polygonum bistorta	Douglas bistort
Pteridium aquilinum	Western brackenfern
Rumex acetosella	Sheep sorrel

	Scientific name	Common Name
	Taraxacum officinale	Dandelion
	Tragopogon dubius	Yellow salsify
	Verbascum thapsis	Common mullein
	Viola spp.	Violets
Grasses	Elymus trachycaulus	Slender wheatgrass
	Bromus carinatus	Canada brome
	Bromus tectorum	Columbia brome
	Dactylus glomerata	Orchardgrass
	Elymus glaucus	Blue wildrye
	Festuca ovina	Sheep fescue
	Festuca occidentalis	Western fescue
	Panicum spp.	Panicgrass
	Poa sandbergii	Sandberg bluegrass
	Poa bulbosa	Bulbous bluegrass
	Poa pratensis	Kentucky bluegrass

INDEX